SCONE

COASTING SAILORMAN

CHAFFCUTTER

Henry Alfred Edward Bagshaw in his early forties.
He was known as Captain 'Harry' Bagshaw to
his barging contemporaries. This photograph was taken
in Glasgow when his barge *Scone* was engaged on wartime
Naval duties on the Clyde.

COASTING SAILORMAN

CAPTAIN H.A.E. 'HARRY' BAGSHAW
With additional material by Albert V. Bagshaw

Compiled and edited by Richard Walsh

**A first-hand account
of life and trade aboard Thames sailing barges
between the years 1915 and 1945**

Chaffcutter Books
1998

The Society for Sailing Barge Research

The Society for Spritsail Barge Research was established in 1963 by a band of enthusiasts concerned that the rapid decline and possible extinction of these splendid and historically significant craft would pass largely unrecorded. From the thousands of Thames sailing barges once plying the estuaries of the south-east and beyond, today just a handful survive in active commission, charter parties and business guests replacing the grain, cement and coal cargoes of yesteryear.

Now renamed The Society for Sailing Barge Research, reflecting a broadening interest in other allied types of craft, the Society organises walks, talks and exhibitions and publishes Topsail, a regular treasure chest of sailing barge history profusely illustrated with fascinating photographs of long lost craft and the ports they once served. Members also receive a twice yearly newsletter which highlights the fortunes of those barges which survive, as well as providing further snippits of our maritime heritage as ongoing research yields yet more of that trade, a way of life which from origins going back hundreds, even thousands of years, ceased in 1970 when the *Cambria* carried her last freight under sail alone.

Membership enquiries to Margaret Blackburn,
9 Waterworks Houses, Layer-de-la-Haye, Colchester, Essex, CO2 0EP.

Published by
Chaffcutter Books
on behalf of the
Society for Sailing Barge Research.

(First published as ISBN 0 9532422 0 X Numbered Limited Edition of 500)
ISBN 0 9532422 1 8 General Edition

Chaffcutter Books is the publishing division of
Isla Giatt Limited, 25 Buntingford Road, Puckeridge, Ware, Hertfordshire, SG11 1RT, England.

Printed and bound in Great Britain by
George Over Limited, Somers Road, Rugby CV22 7DH

This book is dedicated by Albert V. Bagshaw to his parents;
Father for his commitment to the life afloat,
Mother for her forbearance.

THANKS

Many people have responded enthusiastically to the requests made for information and pictures. An early decision was taken to include only illustrations of the period covered by the narrative. As a result, most photographs received of *Scone* in her later years as a motor barge, or later still restored to sail, are not reproduced here. Many contemporary pictures are included to reflect the maritime environment within which 'Harry' Bagshaw's story unfolds. Sources are identified in parenthesis after each caption. Pictures are included for their relevance to the text. Some are from poor quality originals, one or two are taken from 9.5mm cine film shot by Captain 'Harry' Bagshaw in the thirties; inevitably these do not reproduce with perfect sharpness and clarity.

My thanks to Albert Bagshaw for his enduring conviction that this story was worth the telling, for his contributions to the text, and for trusting me with precious memorabilia of his father's career. Also to my many colleagues in the Society for Sailing Barge Research who have, as I most gratefully applaud, saved me much time and frustration by sharing their knowledge of the subject, and who have guided me to sources of information that would otherwise have been missed. I must single out Tony Farnham, Chairman of the SSBR, for letting me include many photographs and postcards from his extensive personal archive. Many thanks also go to Captain Bob Childs for reading the proofs and providing explanations for some of the more obscure of Captain Bagshaw's recollections. Others who have contributed include individuals too many to thank by name, and the staff of museums and libraries the length and breadth of the British Isles.

Last but not least, to my wife for putting up with the many vague answers I doubtless gave to her sensible questions when my mind was selfishly elsewhere, down Swin or Channel with 'Harry'. Thank you Mary.

Richard Walsh

CONTENTS

INTRODUCTION

Over the years since 'A Floating Home' by Cyril Ionides was published in 1918, sailing barge literature has exhibited a prolificacy to rival the barge fleet. Though most of the 100 plus titles to encompass the subject are now out of print, whatever it was that fuelled such a remarkable level of interest seems to show no sign of flagging.

Although my personal recollection of such scenes as the auxiliary *Will Everard* unloading at Colchester Hythe is as vivid as the day it happened back in 1965, the realisation that more than a quarter of a century has elapsed since Cambria carried the last commercial cargo under sail might well lead one to conclude that little on the subject can be left to unearth. Not so, for I recall that when carrying out the research for *Kathleen, the Biography of a Sailing Barge*, through dint of patience and resolve I tracked down over fifty photographs of her in trade. With no reason to think that she was anymore photogenic than the rest, one might imagine that there are fifty pictures of every barge, and in the knowledge that some 1,500 such craft plied their trade well into the present century, a total of 75,000 photos might well be out there somewhere!

Photo: Christel clear

Perhaps therefore it is not so surprising that the Thames sailing barges continue to give up their secrets. Here we can read an autobiographical account of a career in the coasting trade under sail and motor, initially as third hand, then as mate and finally, still at a young age, in command. The narrative has been altered as little as possible from the original set down by the author, thereby capturing the social context of the late Captain 'Harry' Bagshaw's time afloat. This story spans just thirty of his working years, the final chapter covering the end of his career at sea, an event which took place more than fifty years ago. His remarkable memory enabled him to weave both the highlights and the workaday monotony of his experiences into this interesting and historically significant first hand insight into his trade.

We learn of freights to war torn France during the earlier years of the century and latterly something of the fascinating role played by the fast dwindling sailing barge fleet in World War II. From Penzance to the Humber and beyond, ports which have prospered to the present day and others of which little trace remains, all played host to *Scone* in her trading years. Here too are recorded the myriad of passages back and forth for coal, with almost timetable regularity and the times laying windbound wondering from where the next meal might come; a rich narrative indeed.

We share the thrill of his first command, a little 'stumpy' barge in which he poked up canals and river navigations, and which despite her diminutive 90 ton capacity provided 'Harry' with a good living. A teetotaller, Captain Bagshaw spent many of the hours and days waiting for a fair tide or wind, fishing, reading or teaching himself the violin.

By the time he took over the large coasting barge *Scone* he had gained a reputation for pushing his barges hard and introducing innovation to rig and sail plan to gain that elusive extra knot, often the difference between winning the next freight or just 'waiting for orders'.

Much of the time such a career provided a tough but humdrum existence. Certainly it was very hard work, for whilst on passage every hour of the day called for commitment and skill to ensure the safety of his charge. From the story which unfolds it is clear that on occasions a sailorman's lot was downright dangerous. Still, survive he did, remaining fit and alert into his retirement, finishing work as berthing master at a Thamesside paper mill near his home town of Gravesend, and eventually departing this life at the age of 79 in 1980. Likewise so did his barge survive, a fine tribute to the skills of the Frindsbury Lower Yard shipwrights, for the *Scone* is still afloat amidst the hardware of a new age as a hospitality venue for city businessmen in the centre of London Docklands, probably alongside the very same quays she once visited when pursuing a very different trade.

Those parts of this story which concern the family and their trips aboard *Scone* were first written down by Captain Bagshaw's younger son Albert. For the sake of continuity they have been embodied as if related by his father.

Richard Walsh
Braughing
1998

FOREWORD

Many of us go about our everyday lives without thinking about the changing world about us, until it affects our own environment. Then suddenly what we took for granted for so long is no more. Then, with the passage of time, new generations come along and begin to show an interest in some of the discarded ways of their forbears.

Captain 'Harry' Bagshaw was inspired to recall working on the sailing barges in his younger days when a new generation of enthusiastic owners began to restore some of these fine old working craft, which he had thought were gone for ever. Although 'Harry' had given up the open-air life of the sea soon after the end of World War II, the memory of all those working years afloat were still as clear to him as if time itself had stood still.

To illustrate the deep inner feeling he had for the old barging way of life which, I might add, stayed with him right up to the time of his death, I recall how he was quick to pick up the lines and features of any barge, so that he could recognise them when little more than a blur on the horizon. "That's so and so." he would state without hesitation. This ability of his was put to a demanding test, many years after he had given up barging for a life ashore. One day when he and I were out for a drive, far in the distance a sailing barge could just be seen head-on, moored alongside a quay. It looked tiny so far off. 'Harry' said "That's so and so." Later we stopped the car close by; I went over to the barge to check, although by then I should have known better than to have doubted my father's keen eye. Sure enough, the old skipper had proved he hadn't lost any of that old skill.

His last command was the *Scone* in which he served for over twenty-one years. *Scone* was just four years old when he took her over. The first six years of his time in her were spent wholly under sail, mainly trading down Channel, as far west as Penzance. Much later, by then with the added power of an engine, she made her way to Scotland for wartime service on the Clyde.

Life for a sailing barge crew in the coasting trade meant more than a nine to five commitment; it was a way of life. Just like the snail, their home went everywhere with them. Not only did the skipper and mate work the barge, sailing it from place to place, but also provided the labour to load and unload the freight at many of their ports of call. Besides keeping up their homes ashore, the barge had to be maintained in tip-top condition for the rigours of trade. In between freights they were either replacing ropes, painting the decks, sweeping the holds, or scrubbing off and tarring round the hull when in a suitable berth. Like the cliché about the Forth bridge, the job was never ending! And for most skippers it was looking after another persons property, without remuneration from the barge's owner.

That was the world into which my brother and I were born. Although neither of us took up the sea as our livelihood, we came to know most of the ways of the coasting trade. When we were old enough, brother Cyril, mother and I would spend the whole of the school holidays aboard the *Scone*. The barge's boat held great attraction for us boys. As soon as the barge was secured alongside anywhere, off we would go, much to mothers concern.

"They'll be alright." father would say in reassurance. We soon got the hang of it. Once we had ended up on a lee shore, we weren't likely to make that mistake again. Those trips between school terms were the only way we could be together as a family, for the barge was always away, like a scouring vulture of the sea, trying to survive on what scattered freights were available to pick up.

This forward has, I hope, set the stage for what follows. It has been a great pleasure for me to be able to add to father's writings from my own recollections of an era long past. I should also take this opportunity to express my thanks to the Society for Sailing Barge Research, which for over thirty years has championed the publication of material about the Thames barges, and in particular Richard Walsh, without whose commitment this story would never have seen the light of day.

Albert V. Bagshaw
Northfleet
1998

CHAPTER 1

The Unexpected Opportunity

"... I remember the many objections mother made against the venture. She went to great lengths to discourage me."

It was March 1915. Just over six months after the world was beginning to be split asunder by the First World War, although its terrible and far reaching significance had yet to be realised.

But war was certainly not on the mind of a tall thin lad of just over fourteen years of age that March morning as he went slowly down hill over the cobbled stones that formed that quaint and narrow thoroughfare called the High Street, in the riverside town of Gravesend in Kent.

Even at that early age, several jobs had come his way, manual jobs which counted for little and interested him even less. Living in the nearby Parish of Northfleet with his parents, they had one of six cottages just behind the public house called The Volunteer. His parents were of little help or guidance to him, for his father had made The Volunteer almost a second home, which may have helped him with his troubles, or so he thought. While mother found it hard to bring up the children, she strove daily to try and make the proverbial ends meet with scant success. So that then was his home life. No doubt, being the eldest of four children things were expected of him. And who should know better about those times, because that lad was none other than indeed, myself.

In those days the bottom of the High Street was full of people, coming and going, for it was from there one caught the passenger ferry over to Tilbury. But that was not for me that morning. Turning right at the bottom I started to make my way towards Bawley Bay by the St.Andrew's Waterside Mission church. Although the wind was cold coming off the water it had been my intention to view the river. But this was not to be so. Destiny was about to take a firm hand, with the result that it was to change my whole way of life.

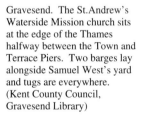

Gravesend. The St.Andrew's Waterside Mission church sits at the edge of the Thames halfway between the Town and Terrace Piers. Two barges lay alongside Samuel West's yard and tugs are everywhere. (Kent County Council, Gravesend Library)

Imagine my surprise then, when a stranger dressed in a navy blue suit complete with bowler hat approached me. His forthright question stopped me in my tracks, "How would you like to go to France on a sailing barge?" The suddenness and unexpectedness of the question stunned me for the moment, especially with its coming out of the blue like that, and from a complete stranger. Within seconds my mind grasped the situation urging me to find out more about the abrupt proposal.

I was soon to discover that my new acquaintance was none other than Mr. Wright, the foreman shipwright of the yard of Samuel West Limited, by the Royal Terrace Pier, and unlike the old time sailing days of Nelson, he was asking me without a press gang, for it was one of their coasting barges that was in need of a boy for a third hand.

In due course I was presented to Captain H. Hoare, master of the Thames sailing barge *Gwynhelen*, who as it turned out was a Faversham man, as was also the mate. By then the idea certainly aroused my interest for this type of life, it could be just what I was looking for. If the skipper was anything to go by, satisfaction just seemed to radiate from his weather beaten face. With the suggestion that a tour of the barge would be to my advantage, we made our way forward towards the fo'c'sle. As we came abreast of the mast I could not help myself from pausing. There, gazing aloft, I saw the freshly varnished spars, with the topmast pointing away skywards. Flying at the top was the house flag of Samuel West. Rectangular in shape, it had both trailing corners rounded. Three horizontal bands of colour made up the flag, the outer two were blue with the centre of orange. I was later to learn it's by that flag or 'bob' that the barge was steered on the wind. My attention was drawn back to the deck level with the Captain now applying some technical barge names. These were too many for me to remember at that time. The only thing that struck home were the pleasant smells. These came from all about me, such as Stockholm tar, new ropes, fresh paints, the sail dressing of cod oil and red ochre. Perhaps it's only those who have been associated with sailing barges that would appreciate these, though I wonder.

Arriving at the fo'c'sle hatch, I followed the Captain over the hatch coaming onto the step ladder. Down below, one had to get accustomed to the half light, which was partly coming in by way of the open hatch and from the two narrow slit deck lights overhead. "Mind the keelson," steering me away as I stepped off the ladder down below, the Captain pointed down to the 9 inch high girder which formed the backbone of the barge. Under the port deck light were two framed hammock cots, at low level were comfortable looking lockers, the buttoned covered cushions of which I was told were filled with horse hair. Removing those gave access to the lockers with lift out type lids, exposing all sorts of things which had been stowed away, like chopped fire wood, coke and coal.

Over on the starboard side was the cooking stove, which I must say looked nice and clean. He must have read my thoughts, "Yes and that's how we like to keep it. It takes plenty of elbow grease." Around the top were rails to keep the pots and pans in place; whatever the angle of sailing they would not slip off. The only means of light when the hatch was closed in rain

[1] *Two massive vertical oak timbers fitted to the floor timbers of the barge, passing up through the fo'c'sle accommodation and deck above to support either end of the anchor windlass.*

'Harry' Bagshaw's National Registration Certificate issued by Northfleet Urban District Council in 1915 listing his occupation as '3rd Hand of Coasting Barge'. (Albert Bagshaw collection)

or darkness were the two oil lamps which swung in their gimbals fixed to the bitts[1]. Underneath and behind the ladder was the fresh water tank. I was to learn later just what it is like to have no running water. Many more items were stowed away in the fo'c'sle, including the different types of brooms and shovels, like the large flat wooden ones used for stowing grain and maize. Back on deck, the Captain was waiting for my decision. "It may sound and look a lot but you'll soon get the hang of it." he said. It was a big step and I began to have doubts upon what action to take for my mind was racing over the problems of home and what they would say. These doubts were soon dismissed by the Captain saying "Why not talk it over with your parents?" I agreed; at least my mother should be consulted. With that I departed.

Even all these years later how well I can remember the many objections mother made against the venture. She went to great lengths to discourage me, finally hesitating, then struggling into her coat in not too good a mood saying "Where did you say that barge was?" By the time I had answered, her hat had been placed squarely and pinned firmly on her head and without saying anymore she was gone through the door.

It seemed ages before mother's return. Finally I heard the door opening and went to greet her. "Alright you can go." she said. It would seem that the Captain had been a very persuasive man. Within a matter of hours my new career started and I was duly signed on as third hand of the Thames sailing barge *Gwynhelen* at the rate of 7/6d per week, food included.

This is to Certify that

(a) Henry Bagshaw

(b) 3rd Hand of Coasting Barge

(c) of 6 Buckingham Road Northfleet

has been Registered under the NATIONAL REGISTRATION ACT, 1915.

Signature of Holder. A. Bagshaw

GOD SAVE THE KING.

(a) Name. (b) Occupation. (c) Postal Address.

CHAPTER 2

Learning
the
Ropes

"Being too excited to sleep, I rose early, for it was going to be my very first day afloat."

When I joined the *Gwynhelen*, she was lying at the barge yard alongside the Royal Terrace Pier at Gravesend. The Pilot's pier; that was where they set out to board the ships on both the inward and outward from the river.

It had been heavy going walking all the way from home loaded down with a large paper parcel of clothes under one arm and two blankets under the other. I struggled aboard, where the mate helped me to stow my gear below. My bunk mattress was a palliasse. Being straw filled it would have the advantage of being replaced quite easily in the event of it getting wet. I found it very comfortable, as the body moulds itself into it making one snug and warm. Once the strangeness of the new surroundings had worn off, I could have slept on a clothes line after all the fresh air and hard work.

It might be as well to run over some of the old *Gwynhelen's* specifications. I say old, because she has long since gone. She was in collision on March 4th 1935 with the steam ship British Prince when bound from Brightlingsea to Deptford with beach ballast. She lay at her owner's Gravesend yard for some years before being broken up. *Gwynhelen* had been built there by Samuel West in 1909 and carried about 150 tons deadweight.

The hull was what they called flush decked, and measured approximately 85 feet long, with a beam of about 20 feet and was rigged in the usual manner of the day, including a large sprit[1] and bowsprit. A pawl type anchor windlass was fitted. It was in fact a sort of double windlass which required a tremendous amount of back-breaking work and experience to operate; experience which, let me say, could only be learned the hard way.

Arriving on board two days before the barge refit was completed provided the opportunity to familiarise myself with most of the deck fittings, of which some could be looked upon as functional obstacles. Worst of these were the sail travellers beams, called 'horses'. Owing to my enthusiasm to see and learn all that was going on about me and not looking where I was going, I ended up walking into these and cracking my shin bones causing great pain for a while. All the time whilst sailing, either by day or night, they would be in use.

The main horse was made of wood and located in between the steering wheel and the main hatch coaming head[2], extending the full width of the barge to just inside the rails. Having approximately the diameter of 10 inches, it was held down in position by galvanised hooped iron straps. Timber support chocks were used at both ends to raise it above the deck level. Those had shaped horn projections fore and aft which provided a type of bollard when bolted securely to the deck. Sliding on the horse was a heavy iron ring known as the traveller. It looked like an unequal figure eight, with the smaller eye twisted at right angles to the rest. The overall height of the horse was about 15 inches above the deck curvature which made it an ideal bench seat when in harbour, splicing the odd ropes.

[1] *The longest spar on a spritsail barge, extending diagonally from the foot of the mast to the peak of the mainsail, almost 60 feet long on a barge such as Gwynhelen.*

[2] *The hatch coaming heads or headledges form the fore and aft ends of the hatchways, following the transverse camber of the deck, their upper edge curved to match the curvature of the hatch covers.*

4

Going forward there was the fore horse. That was situated just in front of the mast case. Although it was smaller in diameter the function was the same with a small difference. Instead of using chocks, the ends rested on top of the rails, where they were bolted down. The choice of traveller could be either that of heavy wire or a medium sized chain. A loop was taken round the horse before shackling up both ends onto the foresail cringles. Besides these two main shin crackers which had to be stepped over when going fore and aft were six bilge pump heads. These were distributed about the deck at different pick up points, some were tucked away, while others were placed over towards the rails. With the working part of the pump removed after use, the open 6 inch hole was plugged by a tampon cap jammed in with canvas and which stood above the deck about two and a half inches.

Being too excited to sleep, I rose early, for it was going to be my very first day afloat. With our breakfast over and the washing up completed we began singling up on our mooring lines. As soon as we had water under, the barge was put afloat by a Sun tug, which helped us through the temporary Barge Bridge which spanned the river from a point opposite the Clarendon Royal Hotel across to Tilbury at that time. Once clear of the barrier we began setting our sails before casting off from the tug and heading upstream, completing our sail setting as we went. Not knowing if I was standing on my head or heels, I either pulled on this or that, or wound on this or that. The leeboard winches were one thing, but the barrels of the mastcase winches were another, but I was soon told how to surge the lines, rather than let them ride up over one another.

Being a young lad and never having experienced anything like that before, it all seemed exciting and mysterious. There was so much to learn. "Stand-by

The floating 'Barge Bridge' which crossed the Thames from the Tilbury shore to the Clarendon Royal Hotel riverside garden during World War I. A large 'boomie' barge is laying at Samuel West's Terrace Pier Wharf. (Kent County Council, Gravesend Library)

on the bowline[1]." was the next order, for we were shortly to come about on a different tack. Then I had to attend to the foresail. Taking up my post by the starboard middle shroud I held my hand ready for the order "Lee-ho." Then as the barge's head was almost halfway through her turn "Let go." was the cry. No sooner had I let the bowline go off the cleat, the foresail went flying across over the hatches and deck to the other end of the horse, setting itself for the new tack.

Following it over I made fast the port bowline by passing it through the cringle on the sail then back onto the cleat on the shroud. It hadn't taken me long to realise what a full man-size job that third hand position could be, and how much was expected of me. Few of my duties may sound formidable now as I list them, such as light the fo'c'sle fire, prepare and cook the food, wash up afterwards. Then there was the after cabin to keep clean, swabbing the decks, cleaning and trimming the oil lamps, for below as well as the navigation lamps, not forgetting the little bits of brass work to be cleaned.

Carrying the last of the flood tide we had to bring up[2] at Belvedere for the night. Nearby was a fish manure wharf. Oh what a smell, it was awful. My memory of that first night always reminded me to pass straight through that area underway in the years which followed. The next flood tide enabled us to reach North Woolwich for we were on our way up to Battersea to obtain a freight of crucibles from the Morgan Crucible Works for Boulogne.

The Duncan and her two sisters amassed around 150 years service between them, operating the Woolwich Free Ferry across the Thames. The loading ramp gantry and dolphins can be seen on the north bank and a 'stumpy' barge carries the ebb down river on an almost windless day.
(Greenwich Council,
Local History Library)

Once past the Free Ferry we had to swing the barge round and catch a turn[3]. From hereon we were going to be towed up under some dozen bridges, with each one having it's very own character and interest for me, the newcomer. Before that could happen though, all the gear had to be lowered. Lowering the gear is the collective name given to the task of lowering the spars down together with the sails. That takes quite a bit of skillful work, which only experience can perfect. Even at that early age of mine I was beginning to see that freights obtained above the bridges could involve quite a lot of extra work, with all that business of upping and downing of the gear. To the best of my recollection the freight charge for that trip had been agreed at £120 lump sum.

CHAPTER 3

The Boulogne Run

"... every now and again her bows went crashing through the waves which sent clouds of water and spray flying..."

Loading completed, next came the task of covering up. The replacing of the hatches and the handling of the hatch cloths was a two handed job. To batten down ready for sea, we had to pull the cloths down behind the battens laying in the hooks fixed to the coamings. While one pulled, the other slid the oak wedges in behind the hooks to hold the batten and the cloth lightly, the final driving home of the wedges was carried out later when all the wedges had been entered. Covering up was certainly not for a windy day, but as one couldn't choose the time or weather, it just had to be done.

With the deck cleared of hatches, next came the preparation for our return run. Down came the gear again. It had been raised on our arrival for the loading of our freight. Finally the tug arrived and we got under tow. With the tug belching black smoke we passed back under the bridges. I found myself looking beyond them to the surrounding buildings which made it as interesting as the run up. Once through Tower Bridge we caught a turn before sweating and straining, re-rigging the barge. From North Woolwich we set sail for Gravesend, where on arrival we brought up, making fast to a mooring buoy.

With only the foresail to stow we were almost done. The sail dropped once the halyard had been released. That allowed the sail hanks to bunch up against the oak wedge which was seized on the forestay just above the top stayfall[1] block. The tack was then unhooked, allowing the sail to be rolled inwards from the foot and leach. Taking hold of the downhaul line, it was pulled down tight before being passed round the sail in a corkscrew type lashing. With each turn it was pulled in tight, finally ending up made fast to the sheet end on the traveller wire, which still remained fastened round the fore horse. With a few heaves on the halyard the whole bundle of sail hung hoisted up to about half the normal hoist, freeing the deck area below and entry to the fo'c'sle.

Our main reason for putting in to Gravesend was to take on provisions for the passage out to sea. Laying off on the buoy meant we would have to use the barge's boat which had a length of about 14 feet. It may be some boys dream to handle one of these, but it nearly broke this one's heart learning.

The pulling of a heavy boat over a strong running tidal current with three people aboard was no picnic. Being long before outboard motors were to arrive on the scene us watermen had to be contented with our only means of propulsion, which was by 'Arms Strong Patent' or in other words oar power. The oars could be either held in position by the use of metal rowlocks, or wooden thole pins when rowing. Another way which was quite a favourite with bargemen was sculling. That latter version of power had some advantage over rowing. Sculling had to be carried out in the standing position with the oar resting in a semi-circular cut out recess located in the centre top edge of the transom. It was usual to have a longer and stronger type of oar for that purpose.

[1] *A very long flexible wire rove through a pair of triple (on some larger barges triple and quadruple) sheave blocks, one at the stemhead, one at the lower end of the forestay, to create a tackle by which the mast was raised and lowered.*

It was held at an angle which one found by trial and error, according to an individual's own height.

It was just as bad to have it too steep as it was to have it too shallow. Speaking for myself at that time, I found it a hard tool to master, more so when a wash was thrown up by passing craft which sent the boat tossing about. `That made it hard to hold one's balance while still trying to make those cross over figure of eight shapes sweep under water. It was essential to hold the rhythmic movements, if you didn't then more often than not one could find the oar blade breaking surface, which not only lost way of the boat's progress but could have the person fumbling about trying to regain balance and control. The main advantage of sculling with the oar over the stern was that the boat could carry a full load of people or gear. Navigation at close quarters was certainly made easier, having no oars jutting out from the side of the boat.

On the subject of swells, perhaps one of these would come along just at the wrong moment, like when one was trying to clamber over the side getting in, which personally I found was most awkward. Stepping out, one could gauge the boats lift on the top of the swell much better, gaining that extra height to step aboard. A light on[1] barge could always drop its leeboard fan down a bit for an intermediate step for the visitor or landlubber to come aboard. My first trip ashore was made along with the mate. He placed me to starboard on the for'ard thwart, while he took up his station to port of the centre thwart. After two days of several runs ashore for stores, my boat handling ability had improved enough for me to make the last trips alone, which gave me great satisfaction.

[1] *Light on and light laden described a vessel without a cargo.*

Samuel West's fine coaster *Gwynhelen* deep laden under full sail off Walton.
(National Maritime Museum, Greenwich, London)

[1] *Lashed tightly to prevent movement*

The Chapman lighthouse marked the edge of the Chapman Sand off Canvey Island. (Conway Photo Library)

[2] *Hinged metal bars engaging with teeth on the windlass to prevent it turning the wrong way.*

[3] *The chain moved across the windlass barrel as it was retrieved. Fleeting moved it back to its starting point so that further chain could be wound in. This procedure would take place many times before the anchor was at the stem.*

[4] *Crab winches, so called for their crab like appearance, were used to raise and lower the leeboards, which were fan shaped boards lowered to leeward to reduce leeway when sailing to windward.*

With the davit falls hooked up the boat was hoisted clear of the water, coming to rest griped[1] to just above the quarter board. We were ready for sea. We let go the mooring on the half flood. With a strong westerly wind at our heels we soon made fast headway rounding the Ovens buoy into the Lower Hope reach. The old Curtis & Harvey factory at Cliffe away on our starboard hand was a scene full of activity at that time. Although we were stemming the last of a flood tide I thought our progress was good. The skipper had noticed how I was trying to judge the speed of the barge for myself, first by looking aft at the wake, then over the side. My inquisitiveness prompted some advice. "Landwards lad, that's where you look. Take a fix on something ashore and then see how long we are passing it; that's known as going over the ground." With the water becoming greener and greener in appearance we rounded Lower Hope Point into the top of Sea Reach.

Being almost high water it certainly looked a vast width of river, with the banks of both the Kent and Essex coast beginning to fan out forming the wide estuary mouth. But that was not so at low water, because it was then that the Blyth Sand made its mark, running well off from the Kent side, with navigation channel buoys, the road signs of the sea, to mark its edge. As we strode on down Sea Reach towards the Swatch, I noticed the red painted iron built Chapman lighthouse over on the Canvey Island side. It stood on well spaced legs which inclined inwards towards the centre body section of the main structure where was housed the keepers' quarters and lantern space. The tube-like legs were braced together by a lattice work of girders. At night its light beam marked the main channel, and it made a good marker during the hours of daylight. On many an occasion when we were passing nearby, it was quite common for the men on watch to give us a friendly wave.

On we drove past the entrance to the Medway and down into the East Swale to anchor for the night. Next morning the wind was still blowing fresh from the south-west. We prepared to get under way by shortening up on the anchor chain. That meant sliding on the cranked iron handles. With one up and the other down we started winding away on the windlass, accompanied by the sound of the pawls[2]. Clink, clink, clink it sounded with every turn we made. Slowly the retrieved chain would drop onto the deck with the links piling up on one another until we would have to stop and fleet[3] the chain and clear it away from the barrel area.

By the time the anchor was up I was puffed. In next to no time we were bowling along under way with Margate on our starboard beam. Shaping a course to round the North Foreland, before we made the Longnose Buoy the mate called for my assistance to help him haul the heavy tackle, for the vang and the main sheet had to come in. The 3 inch ropes were more than the hands of a mere 15 year old lad could handle, not only were they tender but also very cold, making it awkward to grip and pull at the same time. With the big sail thrashing and flapping like a live thing I would have been held in a grip of fear had there been nothing else to do. There I was doubling over the surging deck to the crab winches[4] and then back to the sail with the shouts of the Captain constantly ringing in my ears.

Gwynhelen's Captain was a proper sailing master, fair and good to his boys but very mean with the food rations, or at least that's how it seemed to me, being

The Weymouth *May* deep laden in Weymouth Bay. The large number of 'undressed' patches on her plain sails suggest a recent refit.
(National Maritime Museum, Greenwich, London)

[1] *A minority of sailing barges were built of iron, and from the late 19th C. steel built barges joined the fleet. They were known collectively as ironpots.*

a growing lad together with all the fresh air and exercise. Our few and brief periods of rest were hardly spent in a comfortable manner, because our only means of heating to combat the cold was by burning coke on our stoves. The problem with that was the back wind from the sails blew the sulphurous fumes from the chimney back down into the fo'c'sle, which soon developed an almost unbearable atmosphere. Still, it was just the same aft for the skipper; there it was the wind from the large mizzen sail that blew the acrid fumes down into his cabin.

Finally, we arrived off the Brake Sand and had to bring up at our anchor, for the guard ship had closed the road. We were not alone I might add, for the Weymouth *May* and *Beatrice Maud* were about, together with several of Goldsmith's ironpots[1].

Waiting for the sea road to re-open gave us an uncomfortable time. We were pitching and wallowing about owing to the heavy swell running. After the third day the wind shifted round to the south. Our Captain called us to get under way. He told us he didn't like the look of it, with the barometer glass reading so low, and we would be better off running back to the Gore Channel. That decision was most welcomed by myself, for my stomach had taken enough punishment during those last few days, being sea sick for most of the time. Once under way again I felt much better; having something to do no doubt helped. We certainly didn't leave too soon. As it was, we only just managed to fetch up to our anchorage off the Hook Beacon by nightfall.

By 8 pm that late March evening the wind went into the north-east and blew to gale force. We lay fairly easily with both anchors down and plenty of cable out, but I couldn't help being distracted by the strange noises, resulting in a sleepless night.

Others were not so fortunate as us, for no less than thirteen barges went ashore during that night. Among them was the *Lord Dufferin*, laden with scrap iron for Dieppe, which stranded at Kingsgate never to come off again, while several of the ironpots were blown up on the shore at Margate and Birchington beaches.

When daylight came next morning, while in the process of setting our sails, chunks of ice fell from aloft hitting the deck and breaking into tiny fragments. Our course was back to the Swale where we came across almost fifty cargo craft at anchor in the East Swale alone, all awaiting a fair wind down Channel.

I was to get to know that Swale anchorage, often left quite alone on board the barge, miles from anywhere for days on end. I had no means of getting off because both the skipper and mate came from Faversham and they went ashore together with the boat.

That ordeal was to happen several times to me while I was with the *Gwynhelen;* time just dragged by. There was only one consolation; each occasion had its different weather patterns which helped to break the monotonous hours of waiting. I remember the time when the ice came drifting down, hitting the stem, then as it passed down the sides making several little tapping noises along the hull. There were quieter times when the anchor chain could be heard dragging over the ground as the barge swung round on the change of tide. It was also here that I first heard the cry of the curlews; in fact at that first hearing they really scared me.

In due course, with fresh supplies aboard, we were under way again passing through the Downs and were soon abeam of Dover. Then it was on down past Folkstone, passing through the gate[1], from where we headed across the Channel towards Boulogne. We were passed by several troopships with escorting destroyers, and by the huge swells they sent up they must have been steaming at full speed. When those swells came they often bounded straight over us with clouds of spray, as the water temporarily buried the barge.

[1] The 'gate' was the term given to the northern entrance to the swept (of mines) channel across to the north French coast. Vessels bound over would gather there to await clearance from the Naval craft on station before making passage.

Passing the breakwater on the way into Boulogne harbour.
(Tony Farnham collection)

Standing in towards the French coast we caught our first glimpse of Boulogne with its ancient cathedral dome and the top of Napoleon's monument. By the time we entered port it was quite dark. Surprisingly, we found help was on hand to guide and berth us. Next morning we began to understand why our reception had been arranged. The port was full of activity, with all sorts of craft including quite a number of sailing barges. But it was the troopships that caught my young eyes. They were filled with soldiers, many of whom were singing lustily 'It's a long way to Tipperary'. But alas, there's another aspect to war, for we were to witness a great number of wounded soldiers in solemn mood waiting patiently on the quayside for a ship back to 'Blighty'.

After a day or so in dock lying under the head of a large 'Grange'[2] boat loaded with frozen meat, we commenced discharging with our own gear. That meant heaving the heavy cases of crucibles out of the hold with the windlass, which was a slow and hard job. Finally the last case was ashore and we were making ready for sea again, for we were to make a light passage back to the Thames, up to Greenwich Gas Works. I well remember that freezing cold morning making our preparations to sail, for our wet cotton lines seemed to burn the hands that grasped them. After passing through the lock we tied up at a pier and hove up our boat into the davits. We pulled out the topsail sheet, then the

[2] This 'boat' was probably one of Houlder Bros. ships which had names such as Hornby Grange and Denby Grange.

11

halyard had to be pulled right up to set the headstick over on the starboard side of the mast.

Next we set the mainsail, and got the main tack down. The foresail was placed to windward on a bowline, then set. The leeward leeboard was run down, and before letting go the breast rope, the fore tack was set up tight. Finally away went the spring from its position on the fore horse and with the gap now widening between us and the pier, the line was hauled back aboard.

Our attention was then turned to the anchor, which until then had been hanging just under the forefoot[1]. Being the *Gwynhelen* had a square forefoot we invariably had to use our barge boat's mast as a lever to clear the anchor away from under, while one of us wound on the windlass to bring it right up. On that

occasion, as on many others, that was quite a feat, for by that time the barge was rolling and pitching in the sea outside the harbour. However, after a struggle we finally managed to get it, after which we secured the preventer chain round one of the flukes, pulling the anchor up sideways to lay on the bow, the other end of the chain made fast on a cleat inside the rail. As a precaution we applied the 'dog'. That was a two prong forged iron claw which was made to fit over the chain link edgeways and attached to the deck on a short chain. It stopped the unwanted run out. There was no rest for us, for the bowsprit had to be lowered so we could set the jib sail. By that time we were well off in the heavy swells, the barge tearing through the water, making the task very unpleasant. Every now and again our bows went crashing through the waves which sent sheets of water and spray flying aboard soaking us as well as the already well wetted deck.

A pass allowing 'Harry' ashore in wartime Boulogne, issued by the Intelligence Office. Note that Men had to return to their ships by 9.30pm, half an hour earlier than Officers.
(Albert Bagshaw collection)

The barge was sailing on the port tack. The weather bowsprit shroud had to be held fast initially, and then slacked a little. Then we hauled down the bobstay[2] and set up tight the weather shroud, followed by the lee one. After setting up the jibstay and making fast the foot ropes, we finally loosened the jibsail by letting go the downhaul, and with the sheet pulled in the sail was set. Down below, after getting dried out, it was time to get a meal going. Being closed in gave me another sudden bout of sea sickness, which fortunately I only had to suffer for a few hours. In fact we were only about four hours crossing the channel, and within two days of leaving Boulogne we were waiting at Greenwich for our turn to load a freight of coke for Calais.

[2] *A chain led from the outboard end of the bowsprit via a sheave on the lower part of the stempost and back up to the deck, used to prevent the bowsprit lifting when under load. It was slacked away when the bowsprit was topped up when sailing in the river, or when alongside docks and wharves.*

CHAPTER 4

Farewell
Gwynhelen

"... no sooner had I hit the water, the slack of the line had been taken up pulling me back to the barge."

The Samuel West fleet was quite large in those days. It included the ketch barges *Olympia, Nell Jess* and *Teresa,* the latter two converted to boomies[1] and *Teresa* eventually being reduced to a powder hulk.

Also during my early years with the firm there was the *Lady Gwynfred, Saxon, Gwynronald, Edith & Hilda, Kardomah, Madrali, Nell Gwyn, Rongwynald* which was lost at Boulogne, *H.M.W., T.T.H., British Oak, Abner, Landfield, Madge, Atom* and *Our Boys.* Later on West's bought the *Glenway, Leonard Piper, Iverna* and *Lady Rosebery.* The *Lady Rosebery* was finally lost at Dunkirk in 1940 during the evacuation.

While serving on *Gwynhelen* we made several passages across the channel to the French ports. We were at Calais when the town and harbour were being shelled continually by the German artillery. Then there was the time we were at Boulogne when the big grain ship, the SS Araby sank and completely blocked the harbour.

Gwynhelen at Rye.
(Tony Farnham collection)

Being war time, we had to comply with the extra Board of Trade Regulations for safety at sea. We had three cork life jackets in the barge boat in case of emergencies, together with such like rockets and matches in a waterproof container, a fresh water beaker, a hand bowl in case there was a need to bale if the boat got damaged. We also had on board extra life jackets. These were usually kept underneath our pillows in the bunks. Strange as it may seem, I don't think many bargemen ever thought of putting one on, and because of that one seldom saw them around.

Although there were many like myself who could not swim a stroke, there were indeed instances during my time afloat when I found myself in the water. Luckily for me I remembered what one of my old skippers had taught me, "Boy, keep one hand for yourself and the other for the master." Little did I know it then, but within a few years those words of wisdom were going to save my life not just once, but twice.

Another trip while serving in the *Gwynhelen* which is embedded in my memory was the time that we carried wheat from the Millwall Docks to Dover. We had a fine passage down the Thames and round into the Channel and there

The five-masted Preussen, pride of the German Laeisz fleet, was in collision with the SS Brighton on 6th November 1910. She was towed to Dover where she grounded and became a total loss. This picture shows her some five years later when 'Harry' first saw Preussen from *Gwynhelen*. (National Maritime Museum, Greenwich, London)

we saw the remains of the fine sailing ship Preussen which was still under the South Foreland. On reaching Dover we were towed into harbour by the tug Lady Brassey. Once we were alongside and tied up I had plenty to do. My working day started at 6.30am in order to get the fires lit, for that was our only means of cooking. Having cooked and served breakfast, I cleared away and washed up. That all had to be done quickly because I was expected to help with the uncovering of the hatch cloths and with the removal of the hatches, which would be stacked on the deck around the holds.

When the shore gang was ready it was my turn to take up position on the dolly[1] winch, heaving out the cargo, which was loaded into carts. As the morning wore on, after a cart was loaded and whilst awaiting the next, I took the opportunity to prepare the vegetables and cook the dinner.

In the moments between the loadings in the afternoon, I was busy cleaning and trimming the lamps, sweeping out the cabins and any other jobs the captain wanted doing. At the end of the day we had to replace the hatches and cloths before washing down the decks. Wooden barges needed washing down at least once a day to keep the decks tight.

With completion of our discharge at Dover, we went round to Folkestone to load army huts for Boulogne. We brought up for the night just outside the harbour with about 25 fathoms of chain out. Next morning when we were about ready to enter the harbour we had the devil's own job to get the anchor up.

The general practice was to start setting the sails as soon as the 15 fathom shackle came in board. With the barge being head to wind the topsail was sheeted first, followed by the mainsail. Unless started properly, the setting of a mainsail in a strong wind could be a difficult job. After letting go the peaks[2], then easing the lowers and middles a little, the main brails were let out to the sprit. Then the mainsail tack was made tight. Next the running blocks were overhauled, pulling out the 30 fathom sheet through the double sheave

[1] *The dolly was a general purpose winch fitted to the bitts above the anchor windlass barrel. It had a long light line attached and was useful to assist moving the barge around in the docks. It could also be repositioned aft of the mainmast to operate what was known as the Burton tackle, which was fitted to the sprit and used to work cargo in and out of the main hold.*

[2] *Peaks, uppers, main, middles and lowers were names given to various ropes or wires used to brail (furl) the mainsail (and mizzen) like a theatre curtain.*

[1] *Mousing: A lashing across the jaws of a hook.*

mainsheet block that had to be hooked on the main horse traveller and moused[1] to hold the hook in place.

On that particular occasion we held the traveller ring fast in the middle of the barge by the use of the vang falls ends. The sheet was hauled in on the crab winch barrel as the rest of the middles and lowers were let go and the main brails eased out. It seemed an endless job requiring a lot of energy and muscle power. Once the mainsail was set, we went for'ard to set and hold the foresail to windward on the bowline. To break the anchor away the barge had to be on the right tack. Just when we were ready to start getting the anchor again the chain began to run out, slipping over the barrel. A fourth turn of chain had to be made round the barrel of the windlass, which believe me was no easy job in that blow.

Meantime, our canvas was thrashing about in the wind, with the barge taking turns in sailing in to her chain on one sheer and then the other, each time bringing up with a snatch when she ran out of scope. Whilst all that was going on, we took the chance to get a few turns on the windlass each time our chain went slack. Up came a few links, in went the dogs, then bang as the load came on. As soon as it began to slacken, off came the dogs and a few more links were hove in and so on.

It was probably a couple of hours before we finally got our anchor up and were able to sail into Folkestone harbour. It was only a small drying harbour in which all was well with westerly winds, but you didn't lay so comfortably on the high water with the easterlies blowing.

After eighteen months with the *Gwynhelen* I came to the conclusion that it was time to break away and have a change. After all, my wages had only risen to 10/- per week with food, which was far from being good and so I left.

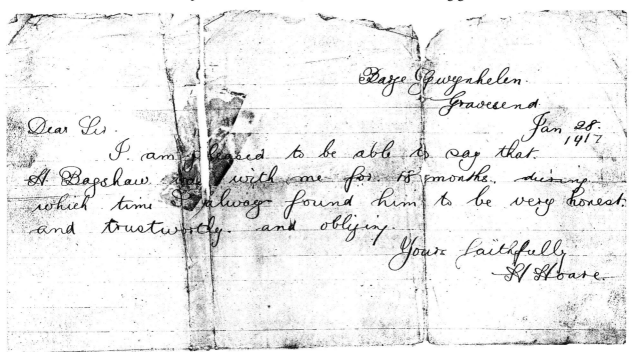

The reference provided by Captain Hoare to 'Harry' Bagshaw when he left *Gwynhelen*. The letter has a full stop after 'honest.'; perhaps after further thought 'and trustworthy.' was added, and eventually 'and obliging.' (Albert Bagshaw collection)

CHAPTER 5

**River
Trading**

"Soon the wind was whistling through the rigging, and by my reckoning it could not be far from gale force."

At 16 years of age I went as mate of a small steam tug. In fact I shipped in two, the Lord Napier and the Invicta, belonging to the Associated Portland Cement Manufacturers. We were engaged in towing lighters between Northfleet, Grays, Swanscombe and Tilbury.

However my life aboard tugs did not last very long, for I had the misfortune to trip over one of the steering chains on deck. Stumbling forward I jammed the middle two fingers of my right hand in between the tug and our tow of barges. It took me almost twelve months to recuperate, having lost one finger top down to the first joint, while the other was badly disfigured. I remember my mother coming to a London court with me with regards to compensation. Although it didn't amount to much, I wasn't allowed to receive it until my twenty-first birthday.

It was the little *Edwin Emily*, another barge, which was my next choice. She was laying at Northfleet awaiting a mate. That seemed to be just the opportunity for me to regain my sea legs. With my gear aboard we caught the last of the flood tide up to White's Cement Works at Swanscombe. On arrival we lay just off the jetty.

After a while the skipper got a little impatient, "Come on, I think we will start unbattening." he said. Round we went knocking out the hatch wedges and collecting them up in a sack, then we lifted out the battens. By that time the sky had become so overcast that we did not dare risk removing the hatch cloths for fear of getting the hold wet. So around the edges we placed all our odd oars, setting booms and boat hooks, hoping to hold the cloths in place. There was nothing else we could do short of putting all the battens and wedges back and that the skipper showed no inclination to do.

He went aft and I went down the fo'c'sle. Later a strong wind sprang up; soon there was a clattering going on up on deck. Climbing up the ladder, I popped my head out of the hatch and was just in time to see the main hatch cloth belly up and go sailing back towards the skipper's cabin hatchway. And for a time that's were it stayed, temporarily imprisoning him. Running aft with my fingers still a little sore from my accident, I clawed away trying to free him but without much success. Luckily for us help was not far away in the form of a passing waterman who had seen our plight and came over to help. Later we had a good laugh about it, for what time and effort had we saved in the end? Quite the opposite I would imagine.

We carried out two or three more freights of cement on the lower reaches of the Thames before sailing round into the river Medway. Our next orders were going to take us under the Rochester Bridge up to the Trechmann Weekes Cement wharf at Cuxton. On our way up river the skipper had been enlightening me as to the way we were to sail under the bridge, without a tow of any sort. That was to be an experience which gave me great satisfaction and pleasure. It was a privilege to have the opportunity to sail with a master with the skills and judgement to undertake such a tricky manoeuvre, albeit with some extra help.

Imagine being aboard a sailing barge, all her sails set, approaching the arches of the bridge, the height of which was not sufficient to give us clearance under. In those days it was the practice to pick up an extra hand known as a huffler[1]. Once we were spotted he sculled out into the stream to meet us. With his boat placed in a good position to allow for our way through the water, he brought her alongside and threw up his long painter for me to catch. After he had stepped aboard, the boat was secured so it dropped astern. Our particular huffler on that occasion had sailed with my skipper before. Knowing what was wanted of him, the huffler took over the wheel to steer the barge towards the bridge arch. Although on some barges the huffler did the lowering of the gear, it was felt that my skipper knew more about how his own barge's rigging went, and would therefore be better when it came to the quick actions needed.

Before actually shooting the bridge, the anchor was put in a chain painter with sufficient anchor chain stacked over the windlass just in case there was an emergency.

With the bridge distance closing and everything prepared, we stood motionless as we waited the last few minutes, our ears straining for the words of command from the huffler which were going to set us to work. At last, as the arch almost seemed to loom over us, came the command "Lower away mainsail." Immediately I dropped the head of the topsail down. That was followed by the skipper letting go the short topmast forestay. The stayfall was ready for lowering the mast. In river barges it was invariably kept in a state of readiness, three turns being made on the windlass and these held to the pinion by a small length of line known as the 'third hand'. When the stayfall was put on the windlass ready for lowering, a piece of ironwork known as the 'Norman' was used. The Norman had three prongs which fitted into three holes bored into a welt[2] of the windlass.

The skipper surged the stayfall around the windlass barrel and those tons of spars and sails were low enough to give us clearance just before our way took us under the arch. The stayfall was put on the windlass while we were still shooting the bridge, the end being passed round and around until short. It became rode over as the first turns of the windlass took the strain. The mast began its journey back up while we were still carrying our way out from the far side of the bridge. Once the mast was up again the stopper[3] was put on the stayfall to hold it from slipping. By that time the huffler's work was finished; he collected his fee for the job done and left the barge in his boat.

There were other cement works above Cuxton; the Hilton, Anderson & Brooks works at Halling, where they even had a small dock for loading. Further up was Lees of Halling and the Peters factory which was on the opposite bank. Above them was Burham by the Sandhole, up to which barges were usually towed, and beyond there we would pass through Allington Lock when towed for Maidstone. During my short time with the *Edwin Emily* we made several runs over those peaceful stretches of the Medway carrying cement, coke and lime.

From the *Edwin Emily* I joined the *G.W.* belonging to W.R.Cunis of Woolwich. She was a high sided barge of about 160 tons burden, and mainly

[1] *An old word for pilot, derived from the Dutch 'Hoveller', a person with special knowledge of a particular river or coastline.*

[2] *A timber, one of eight which formed the bearing surface of the windlass barrel. To create the optimum grip for the anchor chain, alternate welts were made of soft and hard wood. In a working barge they were frequently replaced, as once worn the anchor chain was prone to slip.*

[3] *A small hemp line which, by a series of turns around the different parts of the stayfall between the two stem blocks, locked the tackle.*

used in the coal and ballast trades. Certainly I am not likely to forget the old *G.W.* and the trouble she caused me. Once we were carrying a freight of ballast from Grays. After loading we proceeded up river to the South Woolwich buoys; there we were to moor and await orders. The buoy for which we were making already had a lighter hitched to it, so as we stood up to it we took a turn on the bow cleat and another on the lighter's quarter. With the barge still carrying her headway, that had the effect of driving the lighter ahead as well. To complete our mooring up we required a head rope put out onto the buoy, which had been heeling well over with the tidal run and the way of the barge as we came up.

Not waiting to be told when to go, I jumped with considerable haste onto the buoy, grabbing at the large iron eye ring and the lighters mooring rope. To my horror the buoy heeled over the other way pitching me into the water. However, the weight of the two craft pulling together with the tidal run, soon pulled the buoy back upright again. I was hanging onto the mooring rope shaking like a leaf, before scrambling up to regain a foothold and finish my task. Back on deck I thanked my lucky stars and recalled the good advice I had been given previously. The point had been well proven; what would have happened had I not thought to make a grab for the rope first! Mooring buoys always got their due respect from me after that.

Whilst in the Cunis barge *G.W.* 'Harry' had a couple of accidents which could so easily have cost him his life. (Tony Farnham collection, ex Phillip Kershaw)

The next occasion was slightly different; we were going in alongside a wharf which had ballast heaped up on it. The wind was off and the barge seemed to be fetching up just short of the mark. Being young and inexperienced, on impulse I made a grab for the light heaving line and leapt for the wharf. What could not initially be seen from the barge was the narrow wharf edge which did not leave enough standing room. As I landed I realised there was no room for me. Carrying the momentum of my leap, I endeavoured to climb the heap of gravel with my legs going like pistons, but without success. My plight was soon noticed back aboard, where they stood-by on the other end of the line. With each successive step I could feel the gravel running away from under my feet. In just a few seconds it had covered up the narrow ledge and the next thing I knew water was over my head. Unlike my last time in the water, survival depended on those on board. Sure enough, no sooner had I slid from the quay, the slack of the line had been taken up, pulling me back to the barge.

When I became master it was agreed with my mate, who I might add was much older than myself, that in the event of one of us going over the side while sailing, we would use the timber ladder which would have floated well. With that over the side it would give the person in the water something to climb on, while the remaining one on board could be taking quite a time to bring the barge about to make the pick up. The ladder was really intended for those odd berths that did not have one, and it came in handy over the side when we wanted a scrub round and for getting below in the holds. When we were under way it had its place laying on the hatches within easy reach of the wheel. Naturally we had the more orthodox equipment, like the cork lifebelts hanging on the mizzen shrouds with the barge's name and port of registry painted on them. We also had a lighter type for throwing, which had a strong line attached to it. That was kept handy aft. The person at the wheel would have the best chance of throwing it before the barge covered too much ground. It should be remembered that barges didn't have deep rails at that time, even though it was common to see water on the lee-deck, sailing deep laden. We sometimes rigged a life-line when we knew we were in for a rough time, secured fore and aft.

Another time the *G.W.* caused me trouble was when we had anchored off the north shore in St.Clements Reach for the weekend, waiting to load ballast from the old Tilbury Dredging Company's dredger, Diver.

Sculling back to the barge after putting the skipper ashore for the weekend I became aware of a deterioration in the weather. So much so that by the time I was back at the barge there was quite a fresh wind blowing which was making up a nasty swell. Once I had caught a turn on the barge I lashed the oars down on the middle thwart of the boat, for she was in for a tossing. Grabbing hold of the leeboard top edge I clambered aboard and let the boat drop astern. Down below alone, there was the fire to stoke up for a cup of tea before doing anything else. Soon the wind was whistling through the rigging, and by my reckoning it could not be far from gale force. After a while of huddling by the fire, my thoughts turned to my duties and I decided I'd better have a look around to see how things were standing up in the wind. On reaching the deck it didn't take me long to realise she had dragged her anchor. With wind and tide the barge had been driven down stream as far as Grays. Now what could that barely 17 year old lad do then?

Before long the wind moderated a little and I managed to get the barge more or less under control, planning to give her a sheer inshore and some more chain. But then I changed my mind. Why not try and sail her back to the original anchorage? My enthusiasm got the better of my common sense and was to lead me into, and fortunately through, a nightmare situation. Imagine that big old barge with only myself aboard.

I waited for the flood tide before I set enough sail to give her steerage way. After a few turns on the windlass heaving in the anchor, she wanted a correction on the wheel to hold her steady. Once done it was at the double for'ard again to heave in some more, including fleeting the chain clear. With sweat pouring off me, it was another run aft to fetch her head back on course, then back again to take in some more chain, only to find that it had got foul of the anchor fluke. There was no time for wondering what to do next, for the helm needed another steady, then back up for'ard again to try and pull up a bight of anchor chain by using a chain painter around it. It worked and the anchor swung clear.

During the time I'd spent hanging over the bow and heaving on the windlass, the barge was steering wildly all over the river, the wind whistling in the rigging and the leeboards and rudder making no end of a din. At long last I got her under control without any damage, and managed to get back to our original anchorage, stow the sails and drop the anchor. I felt completely exhausted, both physically and in my mind, but my work was not yet done, for looking aft I noticed our boat lying at the end of its painter completely water-logged. It must have been swamped during that hectic journey.

That water filled boat of ours, being heavy and sluggish, took some pulling in against the tide. Once it was alongside the leeboard I hove the painter short and lowered the tail of the board so I could get to the boat's stern ring and shackle on the vang tackle to support her aft. I then set to bailing the boat out. At least the oars were still there lashed down. Having first removed the bulk of the water my next task was to heave the boat on board.

With the main sheet hook securely lodged in the painter eye ring forward, I hove on the leeboard winch, bringing the vang in tight. Then I put the main sheet fall on the mast case winch. Working these two lifts alternately, the boat was eventually hove up on board where the remainder of the water was taken out of her. Putting her back into the water by reversing the procedure, I undid the painter and let the boat drift astern before making her fast. With the satisfaction of having completed that back-breaking job, I hung up the riding light and stumbled off to my bunk more dead than alive for a few hours rest.

On another occasion the *G.W.* had to take a freight of gas coal from William's Hoists at Dagenham round to Maidstone Gas Works. After waiting at Dagenham buoys for the arrival of the coaler, we struck[1] our topmast before taking our cargo aboard as there was always the danger of the cranes fouling it with their jibs as they loaded us. Outward bound for Maidstone we had to pass through the wartime boom defence across Sheerness harbour.

All us sailing barges had to be towed into Sheerness through the boom. We were expected to assemble at the Spit buoy, awaiting our turn for the tow. One of Knight's smart little tugs called Kite undertook that work; she had a most unusual bow shape. After casting off, we made our way up the Medway to Chatham buoys, where we lowered our gear and made ready for our tow up to Maidstone. However, that winter we had an unusual amount of rain and snow which flooded the river and prevented us from going under the bridges. It was over a week before the water level dropped sufficiently for us to reach our destination and land our cargo.

It had been tricky enough getting our deeply laden barge up river under Aylesford Bridge and through Allington locks, but it was to be even trickier still bringing the light barge back down.

We waited our chance to pull her down to Allington locks and once through we had another wait for the water to fall before we could pass under Aylesford Bridge. Dropping down river we made progress towards the bridge by using a line I took ashore. Sculling off in the boat with one end, the skipper paid out almost 60 fathoms before I had reached the river bank and made fast the boat. I found a suitable tree trunk to take a turn with the line. With the down water, she needed little pulling, in fact only the occasional heave to correct her steerage was required while she drifted along. Once the slack taken up, the process would be repeated again, and again, until we were safely through the bridge. Once through we went to the first available berth to await our tow the following day.

One amusing thing took place on the *G.W.* when we had finished discharging coal at the Curtis Harvey's factory at Cliffe. We were ordered to

[1] *To strike a spar is to lower it. This was done to the topmast when lowering down the gear, except when shooting bridges, and to reduce windage in exposed anchorages.*

Built and owned by Shrubsall, the impressive *Valonia* seen moored at Emsworth in Hampshire. (National Maritime Museum, Greenwich, London)

stand off to the dredger which was at work in the Lower Hope and fetch back to Curtis's a load of ballast. As soon as we had water off, we left the berth and went straight over to the dredger. We loaded so quickly that we saved our water and were back in the berth again on the same tide. Next morning the Foreman came along and let off at us for not shifting. Was his face red when we told him that he must unload us first!

A peculiar feature of the W.R.Cunis fleet that set them apart from almost all the other barges was the black colouring of their sails. Their house flag or 'bob' was a plain red rectangle.

Their *Glenmore* was a smart coaster and during the time that George 'Navvy' Brooks was her skipper she was always kept in fine trim. He was one of only six men whom I ever heard went out of their way to praise their owners. The others were Capt. Ketley and Capt. George Battershall who were both skippers in the *Valonia* during her early days. She was owned by Shrubsall and was a lovely craft. Unfortunately her days ended at Dunkirk, but following collision with a steamer in the dock, not by enemy action. Then there was 'Scamper' of the *Thistle* belonging to Covington's and Fred Farman of Erith when he was skipper of the *Invicta* running wheat to Maidstone, and finally Walter Dossher of the Weymouth *May*. The remainder of the W.R.Cunis fleet was the *Teaser* and *Toots*; the *Glendevon*, *Maria* and *Normanhurst*.

I left the old *G.W.* having discovered that the skipper was far from being honest when it came to paying his mate's wages. It was a great pity for we got on well otherwise and he was a good bargeman.

Normanhurst, built by Cunis at Upnor in 1900 and operated by them in trade, seen here in her original 'mulie' guise. She was later fitted with a spritsail mizzen abaft her short spindle steering position. There are at least five people aboard, including one right aft sporting bowler hat and overcoat. (National Maritime Museum, Greenwich, London)

CHAPTER 6

Some
Close Shaves

"... with a gale of wind screaming about me, and with the barge rolling heavily, I had to go up the rigging."

It must have been towards the end of June 1917 when I joined the sailing barge *Morning* as mate under Captain Marsh. She was owned by 'Gaddy' Knight who also had several such craft named *Afternoon*, *Evening*, *Twilight*, *Meridian* and *Delce*. All the running gear of these barges was fitted with hemp ropes.

I personally asked Mr. Knight to fit some stanchion rails round our barge's bow for safety. My request was met by his blank refusal, but they were fitted to most of the coasting barges within a few years. Could it be that my remarks were instrumental in that coming about, I wonder?

We loaded a cargo of pitch at the Johnson's Tar Works above New Hythe for Le Treport, but later these orders were changed to Dieppe. We came away from Rochester and stood down the river until finally bringing up in Saltpan Reach, not far from the wreck of HMS Bulwark which had blown up there. The next day we crossed over to Southend where we received our sailing instructions from the Naval Control. We went round the North Foreland and through the Downs, spending a few days off Deal before crossing the channel. Boulogne had a breakwater

Knight's *Meridian* under topsail and foresail passes astern of the Shrubsall built *Company* in Milton Creek.
(National Maritime Museum, Greenwich, London)

which gave good shelter against the south-westerly winds, but it was of little real assistance against the easterlies. We lay under the breakwater awaiting a southerly wind. When we did come out to move on down, the wind went round to the south-west and drove up a nasty swell which swept over our hatches.

Once out of Boulogne we rounded the whistling buoy and headed westward past Cap d'Alprech, which was a dangerous piece of ground to approach too closely. Further to the west we passed the seaside town of Etaples, with its harbour and shingle beach, and beyond was the square topped lighthouse marking the Baie de l'Authie. Next we passed the entrance of the Somme where a number of monuments stood commemorating William the Conqueror's landing in England in 1066.

Over to the west of the Somme were high cliffs which carried on westward to Dieppe. Beyond the estuary of the Somme we came to the piers of Le Treport harbour. There was a dock inside the harbour which could only be entered on the flood tide, for the harbour dried out at low water. The actual approach to the shore at Dieppe, on the other hand, was fairly deep. In all it was about a week before we got down there, arriving around 4th July. We were in the company of many other barges at Dieppe and it was a while before we were finally unloaded. We left on a fine morning and ran up along the French coast, squared off on the starboard tack. We decided to try for some mackerel. No fancy fishing tackle for us, just a piece of white rag on a hook and line attached to the boathook, and hung over the rail. Just as we were off the estuary of the Somme we were invaded by thousands upon thousands of flies; one could reach out and gather handfuls of them.

Out over on our port side fishing boats were lowering their sails in preparation for a squall. We did not heed the warning. Soon after, the squall suddenly struck us out of the north, bringing sheets of rain. That was one of those times when it was essential to act fast. First we had to gybe quickly and dowse the staysail. Before that could be secured and the topsail brought down, the barge had heeled right over. Water was washing waist deep over the lee deck and the spars that were in the beckets[1] on the lee rigging were lifted out and washed overboard. An 18 foot barge's oar, to be lifted vertically about three feet requires quite an amount of water. By that time the staysail had run up on its own again, and it was blowing a full gale. The weather staysail sheet, of tarred hemp as were all our ropes, had come right over to the leeward side and was under such a load that the tar ran out of it. The boathook I had been fishing with had been snapped in two. Why the barge didn't turn right over I shall never know. The squall eventually blew itself out, and for my part I had learned a valuable lesson. I wondered afterwards if all those flies may just have had some connection with the freak weather that hit us.

We reached Boulogne where we anchored for the night. The next day we were soon under way shaping our course back to the Medway. With a rapid passage back to Rochester, both the skipper and myself said goodbye to the *Morning* and transferred to the *Wyvenhoe*. I do not know what eventually became of the *Morning*. She was a nice little barge, very clean and tight, built of wood of course, unlike the *Wyvenhoe* which was built of steel in 1898. Owned by the London and Rochester Barge Co., the *Wyvenhoe* carried about 130 tons deadweight.

We joined her in the early days of August 1917 at Strood. Although whilst in her we were to have some difficult moments, she was a good barge. What I did not know at the time was that she had drowned her previous mate, John Couchman, a few months before. They had been on passage from Poplar to Faversham with coal, coming to anchor in the evening off Cheney Rock, Sheerness to await the flood tide for the Swale. At the time there was a moderate wind off the land. Within an hour and a half the wind had increased and swung suddenly from sou'sou'east to the north, leaving the *Wyvenhoe* on a lee shore. By midnight a wild gale was blowing and it came on to poor with rain.

[1] *Inter alia, rope or metal eyes on the shrouds for stowage of spars and oars.*

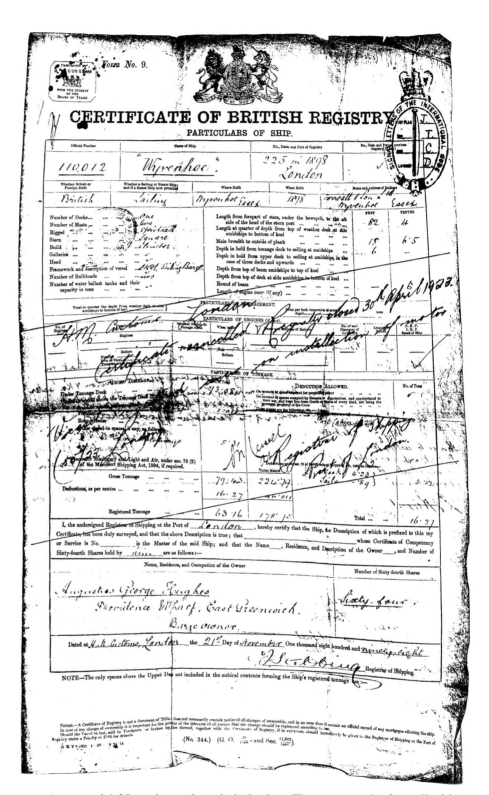

The original registration document for *Wyvenhoe*, endorsed 'Certificate cancelled and Registry closed 30th April 1923. Vessel registered anew on installation of motor'. (Peter Cariss collection)

At around 1.30am the anchor chain broke. The mate set the foresail, skipper James Moore hoping they could make it to Gravesend. But immediately the steering gear broke and they were forced to run the barge ashore on the Sheppey

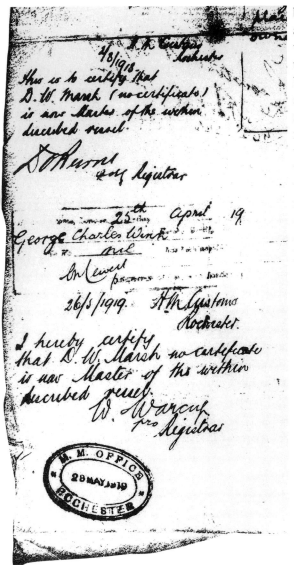

Part of the reverse of the original registration document for *Wyvenhoe*, showing the endorsements for Captain Marsh. He appears to have had a month away from her between 25th April 1919 and 26th May 1919.
(Peter Cariss collection)

beach. That was successfully achieved, but the waves were washing clean over her. When they tried to lower the barge's boat to get ashore, it got smashed on a breakwater and sank. The mate climbed the mizzen and the skipper went up in the main rigging, from which he was washed three times that fateful night, just managing to save himself and clamber back. He shouted encouragement to the mate to hang on, for the tide was ebbing by then and they would shortly have been able to wade ashore. The *Wyvenhoe* was taking a bit of a hammering and had swung broadside to the breakers. A little later the mizzen was carried away taking the unfortunate mate to his doom. It was around 6.00am before an exhausted and numbed Captain Moore struggled ashore, finding the body of his crewman half buried in sand near the stern of the barge.

Despite that very tragic misfortune and the extensive damage to the barge's gear and cargo, the hull was found to be in good order and she was soon refitted and back in service.

For our first freight we sailed up to London where we loaded cotton seed for Hull. Those early days aboard the barge were not quite like the others on joining. Both of us had to get accustomed to her little ways together. From Hull we carried coal across to Calais. That was the first of many trips across the channel with coal which we made in the months that followed, alternating between the ports of Boulogne and St.Valery. It was during one of these runs that I asked my skipper, Dan Marsh, if my younger brother George could accompany us while on his school holidays. It was agreed. So much did he like it, in later years he became the skipper of F.T.Everard's sailing barge *Jane*. He also had the *Beryl* which was renamed *Santille* after she became a barge yacht.

Returning back light to Hull after our first trip, we brought up in the Yarmouth Roads for a while, to replenish our food store. A coasting barge usually had a young lad aboard; indeed that was how I had started out. Taking the third hand ashore with me in the barge's little boat, we beached her at Yarmouth. I left him with instructions, "Keep pulling her up the beach as the tide rises." Later, returning with our rations, I found that the wind had come on-shore and caught our lad napping, resulting in the boat now lying athwart the incoming breakers. With her half full of water she was dragged round and clear of the breaking waves before we hauled her up the beach to bale her right out. We caught our second wind after a short pause and relaunched the boat into a large incoming swell.

Once the stores were aboard we got the barge under way again. We passed Mundesley and Cromer. By the next afternoon the wind veered to the east, and strengthened. Although it was not too bad underway, it became pitch black with a heavy swell making up. During that blow we reduced our sail area to just the foresail. Later the foresail came down on its own. Going forward to see what had happened I got hold of the downhaul and when I pulled it down I found that the head of the sail had torn away from the block and halyard, which was still tight up the mast.

We were a light barge, and off a lee shore, so it was a matter of no small urgency to have it reset. With a gale of wind screaming about me, and with the barge rolling heavily, I had to go up the rigging. After fumbling about in the dark I somehow managed to get the block down. Back on deck, a short piece of chain was found that would make a temporary lashing on the foresail head, so the sail could be shackled back onto the halyard block.

The sail was reset, and it seemed to hold, so we made our way back aft. Only then did I realise how sore my hands really were and back in the light of a lamp I found that my efforts had resulted in several large blisters. Some had broken and were bleeding.

We continued to steer north all night. Soundings with the lead were taken occasionally and when dawn broke we sighted a trawler. The wind had not moderated to any extent, but as the light spread across the sky it brought heavy showers of rain from time to time. It was then that the Spurn lighthouse was sighted and we all drew a sigh of relief as we began our run up to the anchorage. We dropped anchor in the South Roads where, as was the custom in those times, we were to wait for a tow up by tug to Hull.

Even here the wind was blowing hard; so much so, she set back on her chain. The way the chain lead showed she was carrying a lot of weight. After weighing up the state of the tide and the wind strength, the skipper decided it would be wiser for us to lower the topmast to reduce the windage and ease some of the weight off the chain. To do that, the topmast heel rope fall had to be pulled out from its stowage and put onto the heel rope, then hove upon to lift the topmast a little. Climbing aloft once more with my hands still sore from the night before, the pain had to be ignored. Once the topmast was lifted, the fid should have been easily removed. The fid was a square iron bar which passed through the hole in the heel of the spar, and kept the topmast in position, resting on the lower cap of the main mast. For over two hours I struggled with that fid; it just refused to move. That was to be another lesson, for when the mast had been hove up the fore stay had placed itself below the lower masthead cap, and

'Harry' Bagshaw's Identity
and Service Certificate, the
first entry recording his time
aboard *Wyvenhoe*.
(Albert Bagshaw collection)

that had caused the fid to jam. It would have to wait until we got to Hull. I returned to the deck soaked to the skin and wind blown, aching in every limb.

Unable to lower the topmast, the skipper then suggested we should bend our strong tow-rope onto the anchor chain to act as a spring. That involved getting in some of our anchor chain so we could attach the tow-rope, before letting it go again. After more strenuous effort the task was accomplished, and it seemed to do the trick. The wind eventually blew itself out, and the weather improved as we were towed up to Hull. We were soon loaded and our repairs carried out before being towed down again to the South Roads in the Humber's mouth.

The Humber was by no means one of the best rivers on which to navigate; the tide ran very strong and the water was invariably very thick with sand washed from the river bed. Moreover the river was wide open to north-west and south-east winds. Several vessels were lost in the upper Humber through running aground and the tide sweeping over them. Hull Roads was the anchorage for craft waiting for the tide to allow entry into Goole, Selby or Keadby. Barges usually brought up in White Booth Roads in south or north-westerly winds, while some used Grimsby Roads. Larger vessels lay at Hawke Roads, and during easterly winds in the Spurn Gat. The South Roads were sheltered water, but only at low tide.

We lay at anchor in the South Roads, that time preparing for our outward voyage. Everything had to be properly stowed and lashed, and moreover, on a coasting barge every ventilator and stove funnel had to be made watertight. The pumps were checked to make sure they were in working order, and the holds battened down, particular attention given to making sure that the four corners of the hatches were made fully secure. The boat was griped down in the centre of the main hatch and all ropes coiled in place for instant use.

While we were at anchor we fell in with the little steam coaster Moray Firth whose master offered to give us a tow up the coast. That meant we should be in

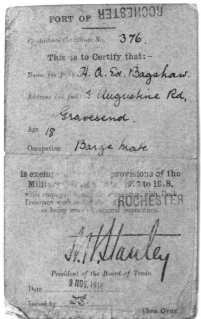

Yarmouth Roads by dark the next day, a run of some eighty miles from Spurn Point.

When we hove up our anchor and passed the tow rope over it was nice and fine, but soon after the tow commenced and we had made the open sea, the wind freshened from the north-west and a heavy swell got up. Quite soon we lost sight of the steamer altogether owing to the height of the seas. We could only tell his position by his smoke. And so we made our way south. At one time we would be lying down in the depth of a trough with the coaster up on a crest, and the next minute we would be up and she down. Suddenly the tow-rope broke. Our steamer turned and came back for us and, in spite of the heavy swell, we gently ran up to him and passed over a second tow rope. It was a truly magnificent piece of seamanship by all concerned. We carried on, reaching Yarmouth by sunset. Thoroughly wet and tired, we all agreed that was enough for one day. The rest of the voyage across to Calais and back passed uneventfully enough.

Once we loaded *Wyvenhoe* a cargo of coke breeze at Beckton Gas Works, again for Calais. After loading we came away and had a fair run down river as far as Herne Bay where we brought up to our anchor, there being no wind. Between eight and ten in the evening as we were looking forward to turning in for the night, a breeze sprung up which developed into a fresh wind from the north-east. We could not turn back as it was a pitch black night; we therefore had to stay where we were. First job was to stow the topsail which meant a climb up the rigging to the masthead, a good 45 feet above the deck, to get the gaskets around the sail. On my return to the deck I was confronted by the skipper who was bleeding profusely and had somehow broken his wrist. I managed to strap up the wound and stop the bleeding.

Then I realised the barge was dragging her anchor. Letting out more chain failed to check her. There was nothing for it but to get the second anchor out, which was housed by the mast case. The anchor weighed between four and five hundredweight and the skipper with his broken wrist was unable to help. We had not even got a third hand on that trip. I had to bend the staysail halyard on the anchor to try and heave it up clear of the deck, then with another line fastened up on a bitt head, haul it forward. While getting it up to the bow I was oblivious to the fact that waves were washing over our decks, drenching me to the skin. Neither was the job done when the anchor arrived up in the bows, for now I had to prepare it for slacking overboard. The stock had to be put on together with the ring and pin, and the shackle for the chain. After uncovering the chain pipe[1], the chain had to be pulled up from the locker below, taking two turns round the windlass, making sure to keep it clear of the main anchor chain.

[1] A second anchor chain locker housed the chain for the spare anchor.

That was hot work in spite of the howling wind and cutting spray. I passed the heavy chain through the hawse pipe, shackled it to the anchor stock and put a lashing on the fluke of the anchor to take a bit of the weight off the chain. Working as if a blind man, for it was an inky black night, I slacked away the staysail halyard a little and took in the chain so I could take off the line which had held the anchor forward. Then I slacked the halyard right up, so the anchor's weight came on the chain. Once certain that everything was prepared on the second anchor, the foresail was made ready to set on a bowline before being run up. The barge surged ahead until the main anchor chain came taught abreast. At that moment I let go the second anchor and laid out the chain. I let the foresail down straight away, but it ran foul of the windlass and the foretack swung round dangerously in the gale. It was eventually captured and made fast so I could get to both anchor chains without being knocked senseless, a situation which would have left the barge without an able bodied man aboard. The chain on both anchors was slackened off. Somehow I managed to stow the foresail, and then checked that the fo'c'sle hatch was secure after covering up the chain pipes to keep out the sea.

I made my way aft over the swaying deck to see how the skipper was getting on, and hoping for a break. It was at least three hours since I went for'ard. It was only then that I had time to find out how he had fractured his wrist. The rudder had been kicking from side to side in the heavy swell. The barge was fitted with chain steering gear and while the skipper was checking to see how it was standing up to the extreme conditions, the back lug of the port block had pulled off as the chain snatched. In an instant the tiller had swung over and hit him, catching his arm between the chain cover and tiller arm.

I had no choice but to attempt a temporary repair to the steering straight away. The rudder was wound hard over on its good side, to starboard. I went for'ard again, opened up the fo'c'sle hatch and went down to get a spare block. After hauling it out and again securing the hatch against the seas, I found the shackle spanner which was kept in the cabin top[1], the *Wyvenhoe* not being a flush deck barge. I undid the broken shackle and removed the damaged block. I shackled the new block on, rove the chain and shackled it to the tiller arm.

At long last everything was in order on the barge, and it was time to think a little of ourselves. The position was nonetheless precarious in the extreme. The only able bodied man aboard the craft was myself, a lad not even 18 years old. It was completely out of the question for the skipper to try and do anything, for he was still in a great deal of pain. We decided to burn flares to bring assistance. While we waited we made some tea and had a meal. The wind showed no sign of moderating and we had barely swallowed the last mouthfuls of our food when we heard another heavy crash right aft. The skipper at once suspected that another block had given way, so with the aid of our stern light I carried out an inspection. My heart sank on having a look, for while the block and chains were still in order the iron rudder post had broken.

Collecting a spare length of chain from up for'ard, an old leeboard pendant[2] in fact, together with a couple of shackles, I set to work unshackling the mizzen sheet block, and fixing a length of chain to extend the short piece that was on the rudder. That I led to the leeboard winch and hove on it, and was

[1] *A raised area above the aft accommodation to provide headroom below. The hatch to the companionway often created a ledge or shelf which was a useful place for tools that might be required in a hurry.*

[2] *A chain connecting the leeboard, via a single whip purchase, to the crab winch.*

E.

CERTIFICATE OF DISCHARGE.

FOR SEAMEN NOT DISCHARGED BEFORE A SUPERINTENDENT OF A
MERCANTILE MARINE OFFICE.

ISSUED BY THE
BOARD OF TRADE
IN PURSUANCE OF
57 & 58 VIC., CH. 60.

Name of Ship and Official Number, Port of Registry and Tonnage.	Horse Power.	Description of Voyage or Employment.
S. B. Wyvenhoe No 110012 London 63 tons	1	Mate

Name of Seaman.	Year of Birth.	Place of Birth.	Capacity.	CERTIFICATE (if any). Grade.	Number.
H. Bagshaw	1900	Gravesend	Mate	—	—

Date of Engagement.	Place of Engagement.	Date of Discharge.	Place of Discharge.
1 aug 1918	Rochester	29/4/1920	London

J Certify that the above particulars are correct and that the above-named Seaman was discharged accordingly.

Dated this 29th day of April 1920.

Signature of Seaman: *H. Bagshaw*

Signature of Master _D. Marsh_

Signature of Witness _J. W. Adams_

Occupation _mariner_

Address _Friary Place Strood Kent_

(B25687.) Wt. 37158—336. 80,000. 12/18. **Gp. 156a—617.** S. & S., Ltd.

Price 4d. per quire.

NOTE.—Any Person who forges or fraudulently alters any Certificate or Report, or copy of a Report, or who makes use of any Certificate or Report, or copy of a Report, which is forged or altered or does not belong to him, shall for each such offence be deemed guilty of a misdemeanour, and may be fined or imprisoned. N.B.—Should this Certificate come into the possession of any person to whom it does not belong, it should be handed to the Superintendent of the nearest Mercantile Marine Office, or be transmitted to the Registrar General of Shipping and Seamen, Tower Hill, London, E.

Certificate of Discharge recording 'Harry' Bagshaw's departure from *Wyvenhoe*. (Albert Bagshaw collection)

[1] *A chain from the tail of the rudder, led to one quarter and made fast. The kicking chain was tensioned by turning the barge's wheel until the slack was taken out of the steering gear thereby stopping the rudder jarring when at anchor.*

Opposite page: Wyvenhoe's skipper, Captain Marsh, credited 'Harry' with 50% more time than he served in this glowing reference. (Albert Bagshaw collection)

overjoyed to find it acted successfully. By leading the proper kicking chain[1] to one crab winch, and my makeshift version to the other, I had restored control over our rudder. No sign of help in response to our flares was forthcoming during that time, and we laid there all that night, all through the following day and also the following night before a boat put off from the shore and took the injured skipper off.

Under the command of another skipper who had been sent out from Rochester, we sailed her back up the Medway with the crab winches acting in place of the steering gear. We were put on the yard, repaired and completed our voyage with coke to Calais. Then we carried coal from Goole to Saint Valery-sur-Somme and after that we loaded cement at Peters' factory above Rochester Bridge.

There followed another run across to Saint Valery with coal. On that trip we were towed back across the channel to the North Foreland by a little ship called the Active whose skipper, curiously enough, was the same man who gave us the tow up the East Coast, when he was in the Moray Firth.

Although his wrist eventually healed and he became fully fit again, the old skipper never returned to the barge. In all, I was almost a couple

20, Stanhope Rd,
Strood,
Kent.
20/8/21.

Harry Bagshaw has been in my employ for 3 years, and has given me every satisfaction He is a very good worker and timekeeper and is very honest and trustworthy and I cannot keep from recommending him with a very good character. His reason of leaving me was to better himself in a larger vessel.

Capt. D. Marsh.

of years in the old *Wyvenhoe*, serving in her from the summer of 1918, through the closing months of the war, until the end of April 1920. I eventually left her to join the *Crouch Belle* because of recurring troubles with the replacement skipper.

The *Crouch Belle* was a wooden barge of about 150 tons deadweight. Despite serving aboard her for about two years, the only voyage out of the ordinary was one from Rochester to Truro and back which lasted all of four months. It was on the 31st July that we sailed from Strood, having loaded a cargo of oil cake, but we fetched up at Dover with the wind against us. We stayed at Dover for the whole of the month of August while a strong wind blew out of the south-west, often bringing heavy rain. When we finally turned our back on Dover we crept down the coast to the Isle of Wight and brought up in Ryde Roads. Then on to Cowes Roads; then Jack-in-the-Basket, Yarmouth; Isle of Wight Roads and then away down to Portland, one small step at a time. Coming out of Portland we ran down to the Bill, but we were forced back again before finally running down to Plymouth. From Plymouth we went on to Falmouth where we sailed up the delightful river Fal to Truro. I have been up and down the Fal several times since and have never failed to marvel at what a pretty river it is.

At Truro we discharged our oil cake cargo and then went to Fowey to load china clay for Queenborough. Having come out of Fowey and taken advantage of the tide, I turned to carrying out my job of clearing up the deck and so on. As I hove tight on a lashing across our hatches, the rope gave way and went over the side. Fortunately for me the leeboard pendant chain was slack, for as I went overboard I managed to make a grab at it, and after getting my breath, managed to haul myself back on board. It was one of those times when both hands were better than one to save oneself, and surely one more case for the provision of rails around a barge's deck.

From Fowey we beat back into Plymouth and then up to Portland and on to the Wight.

Crouch Belle under sail but
short of a breeze.
(Tony Farnham collection)

First we brought up under Ryde and then moved over to Stokes Bay before
making the next run up to Newhaven. Here we lay for three weeks before
making the final passage up to Queenborough. At the conclusion of that
voyage the barge was put up on the blocks to have her bottom scraped, to
free the mussels and other marine growth which had made its home there
during the weeks she had not taken the ground.

Some time later we were lying under the west shore at Sheerness one night.
The A.P.C.M. barge *Bride* came crashing into us, breaking our wale on the port
quarter, putting us in the yard for repairs which took some weeks to complete.

We made one trip to Yarmouth in addition to several runs to Ipswich, but
much of our time we spent laid up, sometimes for a whole month at a time,
waiting a cargo. During these times all my savings would be drained away in
order to live, and I used to look after the barge as she lay at Gravesend for
nothing. I celebrated my twenty-first birthday during those hard times and, of
the opinion that there was still a good living to be made by anyone willing to
make the effort, I determined to have my own command as soon as the
opportunity arose. Things continued at a pretty low ebb for many months
aboard the *Crouch Belle*. However, the work situation did eventually pick up
and with it an opportunity for my promotion to Captain.

CHAPTER 7

Small Beginnings

[1] *Stumpy was the name given to smaller river barges that had no topmast and topsail. They generally had relatively long sprits and higher peaked mainsails.*

[2] *The 'moulding work' for which the sand was gathered for cast iron manufacture in moulds made by compacting the sand around a wooden pattern, carefully removing the pattern then pouring molten iron into the hollow cavity to create a 'sand casting'.*

The Lee Navigation was a busy waterway serving the industry of north London and the malting and milling centres of east Hertfordshire. Two stumpy rigged river barges lie at malting wharves in Ware. The towing path can be seen to the left of the picture.
(Charles Wilkinson, C & C Photography collection)

"... where we sighted a whole variety of yachts, including Shamrock, White Heather, Britannia and the Royal Yacht, Victoria and Albert."

My first command was the 80 ton river stumpy[1] barge *Camelia*. She was a comfortable little craft, although she really only carried about 70 tons and was by that time already about 70 years old. Craft like those made the ideal induction between thinking you knew it and actually carrying it out.

Our first trip was in early March 1922, bound for Holman's up the Surrey Canal with 45 tons of tarred pavings blocks which we had loaded up above Allington Lock on the Medway. I skippered the *Camelia* for about a year, during which time we were trading mainly from Maidstone to London, where we were usually seen up Bow Creek or the Surrey Canal.

Probably the hardest work we did was in the Leigh sand trade. The sand was used for moulding[2] work and we took several loads away during my time in the *Camelia*. The barge was run up on the sand, and as the tide ran out we would go over the side with what was known as a 'Fly' tool, which was really a kind of wooden spade. While the tide was out we would be hard at work, with the mate one side and myself the other. We would cut out the sand and throw it up 6 feet and more into the barge. Working away from the side of the barge the spade became heavier and heavier, and the throws were getting shorter and shorter as our energy drained away from us. The first throws had made the hold, but the later ones had only made the deck. By the time the tide began to flood it was time to get back aboard and clear the decks while the barge re-floated. We would then shift our anchorage to another spot where the whole operation was repeated. Like that we would load about 70 tons during two tides. Then we would cover up the hatches and sail away.

Sometimes we loaded sand at Aylesford for Tottenham Hale Power House by way of a passage up through Lee Bridge on the River Lee Navigation. We negotiated the bridge without much trouble on the way up, with our leeboards unshipped and taken aboard, the gear lowered down, and our boat stowed on deck. After sailing into Bow Creek we would push, pull or row the barge, or if lucky enough get a tow to Bromley Lock. After passing through, a horse would tow us the remainder of the way from the bank. On arrival the gear had to be hoisted in order to uncover the holds. With the barge empty and high out of the water, the return journey was always going to be a different tale when we came on down to Lee Bridge. Then the fun began as we made our preparations in order to pass under the

arch. First the fore hatches were turned up side down and the hatch cloths re-spread over them so as to hold water. Then the barge's boat was brought around and swung athwart of the stem, and with the help of the leeboard tackles[1], the boat was hauled up onto the bows. It was then filled with water by means of a draw bucket over the side. As the weight of the boat increased it began to set the barge slightly down by the head; by filling the fore hatch with water she would drop even further. That enabled the barge to ease her stem and bitt heads under the bridge.

The barge was fitted with a plug in the bottom, which was then pulled out, slowly filling the barge with water. As soon as she began to go through the plug was replaced. The boat on the bows was then eased down into the water and the inverted fore hatch covering bailed out to bring the barge's head up. Slowly her head began to lift and the water in the hold ran aft, sending the after part of the barge down in the water, enabling the rest of the barge and her gear to pass under the bridge. Once clear, all the water was pumped out. Then we could set about getting her ship-shape again.

In that little barge I earned a good living, but only as a result of almost continuous hard, very hard work. Several times did I sleep the clock right round, after a hard spell. Often we would load at Aylesford, or indeed anywhere up as far as Maidstone, then we would be towed down river by one of Knight's tugs, and then we would rig our gear at Chatham before sailing away, saving a tide out of Sheerness. Sometimes we were not so fortunate and had to bring up. But on one occasion after loading paper at Reed's New Hythe Mill a tug gave us a start down Chatham Reach at about the same time that the stumpy *Jim Wigley* was under tow out of Sheerness, some six miles ahead of us. We cleared Sheerness with that tide and rounded up into the top of Sea Reach by high water.

Twenty-one year old 'Harry' Bagshaw, master of the *Camelia*. (Albert Bagshaw collection)

By low water we were off Gravesend, and here I saw black clouds massing ahead over Northfleet Hope. Instinctively my thoughts went back to that sudden squall experienced off the Somme while in the *Morning*. We began stowing up the mainsail, leaving the foresail and mizzen set. As expected, the rain came followed by much wind, but our little *Camelia* lay quite comfortably under her reduced sail. Meanwhile, I watched other barges downing their foresails and coming to anchor in the sudden heavy squall. Not us however, for we passed up Gravesend Reach overtaking the *Jim Wigley*. There were too many outward bound steamers coming down river for us to turn and anchor so I carried right on up to Blackwall Reach. It was not possible to find an anchorage off Greenwich College, so on we went until, believe it or not, we were right up to the Surrey Dock and were locked in, leaving the *Jim Wigley* many reaches behind.

The life of the poor little *Camelia* was drawing to a close however, for about a year later she was run down while in Woolwich Reach drowning her crew of two. Although by then I was on another barge, her loss did mean a lot to me, for as well as losing some of my fellow workmates, she had also been my first command.

My first large barge was the *Knowles*. She was about 160 tons deadweight, carrying some 145 tons of cargo to sea. She lay at the Frindsbury Barge Yard at Strood when I joined her, owned by the London and Rochester, whose house flag carried the white crescent on a red background.

Our first assignment was in May 1923, to carry timber from the Surrey Commercial Dock to Ramsgate. We loaded the timber in the Station Yard of the Surrey Dock and came out through the Greenland entrance. We were locked out together with other craft belonging to the Thames Steam Company, which were unattended by lightermen, and in so doing we sustained some damage to our stem. That I reported to the 'Runner', a man who kept an eye on craft all round the dock, but was not believed. So I had to go to the lock entrance myself and inspect the lock log book to have my contention verified.

We had a nice run down the river after that, passing all the familiar landmarks. Slipping down Limehouse Reach, past Deptford Victualling Yard and Greenwich College, we wound our way round Blackwall Reach and Bugsby's Hole into Woolwich Reach. On through Galleons Reach with the aid of a westerly wind, we were soon clear of Barking and Half-way Reach and into Erith Rands. It was a steady run through Long Reach passing the old Cornwall laying at Purfleet and then on into St.Clements Reach. The Cornwall had been built largely of teak in 1813 as the Wellesley, a two deck

Below:
The Cornwall was a reformatory ship for boys run by a management committee on behalf of the Home Office. She was moved to Denton in 1926 as her Purfleet berth was in the way of a new coal wharf to be developed there. Before WWII most of the reformatory activities had moved ashore.
(Kent County Council, Gravesend Library)

Below right:
Cornwall after being hit by a bomb during the war. She was broken up shortly afterwards.
(Kent County Council, Gravesend Library)

[1]The Exmouth was a 1904 built steel replica of a 'wooden wall' operated by the Metropolitan Asylums Board as a training ship for pauper boys. She eventually changed her name to Worcester when her predecessor of that name sank at her moorings and she took on the role of the Thames Nautical Training College for Merchant Navy officers.

line-of-battle ship of 74 guns. She had moved to Purfleet to replace the original Cornwall in 1867.

After rounding the point into the Northfleet Hope, where we passed the Exmouth[1] laying off Grays, we entered Gravesend Reach. By the time we were into the Lower Hope the sun had begun to go down astern of us, and as dusk fell we brought up to our anchor for the night and for the tide, at the Yantlet in Sea Reach.

A nice breeze was blowing the next morning, and the day was still young when we were again under sail. With our wake stretching out astern we went on down the Swatch across the Sheerness Channel to the Spile buoy and through the Four Fathom Channel to the West Last. Onward through the Gore, past the Reculvers, Birchington and the sands of Margate to the Longnose buoy.

By the time we had rounded the North Foreland the wind had fallen away and as we made our way through the Old Cudd Channel I wondered if the tide would come up to the eastward too soon. It was better to keep the barge on the move as the wind was so light, so we got the oars shipped and commenced to row. In that way we made the piers and entered Ramsgate harbour on the slack water. Generally speaking, Ramsgate harbour was a very bad place to enter on an easterly tide. As soon as we were in, our timber was discharged and we went back to the Medway light. At Strood I protested about the damage done to the stem, and also lodged a complaint about the poor fit of our sails, which were then altered to my satisfaction.

We were then ordered to Cliffe Cement Works, where we loaded a cargo of cement for Torquay, my first long distance passage in command. I felt pleased that I had asked the previous skipper of the *Knowles*, Jimmy Adams, how best to load her. His opinion was to trim her down about nine inches by the stern. Once the 140 tons of cement was stowed away below we were on our way down to the Swatch where we anchored in order to make ready for the open sea.

The bowsprit was lowered and the jib made ready. Different barges had different methods of setting that sail. Some had a running sail, or traveller, on the bowsprit to pull the tack out, with the halyards on one side of the masthead and a tackle block on the other side to pull the leach rope tight. We had a stay to which the jib was fastened with hanks, and there was a rope downhaul. Although that arrangement made it better for handling, it forced us to stow the sail on the bowsprit.

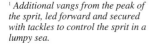

¹ *Additional vangs from the peak of the sprit, led forward and secured with tackles to control the sprit in a lumpy sea.*

Next we hauled up the boat into the davits; ropes were lashed onto the hatches, and believe me we saw to it that there were no rotten ones. With the rolling vang¹ tackles put on, we were ready to face the run down Channel by high water the next morning.

As our craft was drawing about seven feet, I decided that we would not be able to go down 'over the land' as the saying goes, but should take the outside route into the deeper water down to the Girdler light vessel, past the Red Sands

The Girdler light vessel, one of the marks on the 'outside' route, further into the Thames and away from the shallow waters adjacent to the north Kent shore.
(Tony Farnham collection)

and then shape a course in towards the West Last buoy. We followed that course without incident, passing between Margate Sands and the Copperas Bank, before fetching up to our anchor in the Gore to wait a fair tide. The new tide saw us round the North Foreland and into the Downs where we anchored off Deal. The next tide took us along the coast off Dover, past Folkestone, Hythe and Sandgate, to another anchorage in Dungeness West Road.

It was pleasant and the weather was fine, for it was the end of June. Light westerly winds ruffled the surface of the water which sparkled in the strong sunlight, and it was these same westerly airs that compelled us to let go the kedge anchor on the end of a cotton line each time a tide finished its run to the west'ard, in order to avoid us drifting back. Eventually, we were standing into the Fair Light, with the high ground to the east of Hastings. Here the tides are a little different than further east, for at Deal and Dover they run to the east about two and a half hours after high water, while at Hastings the tide makes to the west'ard at about high water time at Dover.

The light westerly breeze was on our faces as we carried on to windward down the coast past St.Leonards and Bexhill. Trying to gain every inch, we had all our sails set, including the flying jib on the bowsprit, as well as the jib. At Pevensey Bay, opposite the Martello tower, we anchored for the night. As soon as the tide began making to the west'ard the following morning we were on our way again, past Eastbourne and down to Beachy Head. As the westerly wind persisted we stood off to the south'ard on the starboard tack. Later, when I estimated that the tide had finished making to the west, we put about to fetch up somewhere on the coast by nightfall. We made our landfall just inside Eastborough Head and brought up to our anchor in the dark under Selsea Bill near Bognor. It was not at all a bad fetch. We had a fine sail on a tight sheet all day, with the water blue and smooth.

From our anchorage we pressed on through the Looe Channel by dark. Down Channel we steered, passing the Mixen Beacon and then on through the Pullar Bank buoys before standing away for the Warner light vessel.

The names of the milestones of our journey would have sounded curious to a landlubber, like the Dean Tail Spit, for example, which we left astern before proceeding in past the Horse Sand Fort on a course set for an anchorage to the west of Ryde pier. By that time I was feeling dog tired, so there we enjoyed a good twenty-four hours rest before heading away down past Wootton Creek, and caught a glimpse of Quarr Abbey on the land as we glided by.

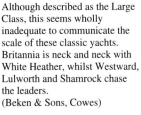

Although described as the Large Class, this seems wholly inadequate to communicate the scale of these classic yachts. Britannia is neck and neck with White Heather, whilst Westward, Lulworth and Shamrock chase the leaders.
(Beken & Sons, Cowes)

With the wind still coming out of the west, we kept inside the Mother Bank, passing Osbourne House to Cowes, where we sighted a whole variety of yachts including Shamrock, White Heather, Britannia and the Royal Yacht, Victoria and Albert.

With Cowes falling astern we pushed on down the Solent where the tides are much stronger than off Spithead. About Yarmouth, Isle of Wight, we fell in

with two other barges, the *Leonard Piper* and the *Glenwood*. Our hope for the next day was to fetch up in Swanage Bay, bearing in mind the continuing westerly airs. However, by the time we had cleared the Needles and the Shingles Sands, the wind died right away leaving the sea dead calm and in next to no time we were enveloped in fog. Close at hand still were the other two barges whose skippers, both of whom were older and much more experienced than myself, sought to drive into the light somewhere. Holding our course we continued, until off our starboard side a light breeze from the east'ard showed on the water. We gybed, bringing the sails over onto the starboard side. Light as it was, we were soon gathering a little way, drawing ahead of the other barges.

The idea of going into Swanage Bay filled me with some apprehension, for I remembered how the *Diana* was lost on the Peveril Ledge. As the wind began to freshen we set our balloon foresail and it was not long before Anvil Point came into sight. With the fog almost gone I got a bearing fix and decided to make the most of the wind, and so shaped a course clear of the Portland Race. With some fifty miles to cover we ploughed on before the freshening breeze from the east, hauling down the balloon foresail before dark, as it then looked as if the fog might return. Bearing in mind the darkness, with a bit of mist about, we blew the fog-horn frequently to warn other craft, especially submarines on the surface, of which there were quite a few.

We continued on the port tack until daylight when we gybed over, having had a nice breeze all through the hours of darkness. It was still foggy enough for us not to see the land, so I cast the lead and found we had nine to ten fathoms of water, so we hauled on to the wind a little. I had been on deck all night with the poor visibility, and had not seen a thing, or heard anything for that matter. Yet, somehow I felt by the hours we had been sailing we should be somewhere near Torquay. I asked the mate to take another cast. Even before his answer came my concern was aroused by the amount of line that had been flying through his hands; mark after mark it went until he cried out fifteen fathoms, which really started me worrying! We hauled the sheets and put the barge about, and within ten minutes it was a relief to find the fog lifting and Berry Head abeam.

The easterly wind died right away, and we spent the rest of that day trying to get into Torquay to anchor. In the early evening a thunderstorm broke out and as the wind came westerly we sailed into Torquay with the tide. Torquay had an inner and outer harbour. The inner harbour dried out and boats were moored in every conceivable place. The outer harbour was good in spite of the fact that a nasty ground swell could be encountered when the wind blew out of the east and south-east. We unloaded our cargo of cement, with myself highly appreciative of Torquay which I decided was a lovely place in summer. But what a difference in the winter time!

My first glimpse of Tor Bay on that occasion had been as the fog lifted, showing the high ground of Berry Head with a landmark on the top of the high cliff. Tor Bay itself was in the shape of a large semi-circle with Berry Head lying to the left hand with Brixham Harbour just inside it, while Torquay lay to the extreme right surrounded by high cliffs and outlying rocks. Paignton was rather nearer to the Torquay end of the semi-circle. The bay was open to the south-east winds, but sheltered against the prevailing westerlies. On a fine day the red soil that backed the bay blended beautifully with the blueish green hue of the sea.

We sailed from Torquay to load our clay cargo at Teignmouth which we were to carry to Queenborough. Teignmouth entrance had a moving bar of sand and shingle. On the approach there was a low shingle beach to the east and a very high cliff to the west. While there I learned that a few years before, the barge *Perseverance* had stranded on the beach when cutting the east side too fine whilst pulling in with a strong easterly wind. The very next swell had caught her however, and she was swept into the centre of the harbour. The harbour entrance took a sharp turn to the east and provided good shelter. Generally vessels discharged at wharves and then moved out to the buoys to load. The clay, which was the principle outward cargo, came down from the pits at Newton Abbott.

It was loading of that cargo which gave me my first experience of not only paying heavily out of our freight money, but working doubly hard into the bargain! The clay came down to us in lighters. That lighterage cost had to be paid out from the barge's hard earned freight money. Once alongside however, we had the task of getting the clay out of the lighters and into the barge. That did not make sense to me, so there and then I resolved not to both pay and do the work again if I could help it. The weather was very hot and by the time we finished loading my hands were rubbed raw.

With our load of clay aboard we left Teignmouth, catching a fair wind that gave us quite a smart passage up to Queenborough. After discharging we went up to Aylesford for a freight of bricks for Newport, Isle of Wight. Before loading we had one or two minor repairs carried out. Although I had found fault with the mast, the spar maker assured me that it was all in good working order, and ought to be considering that the barge was only thirteen years old.

We had a good trip down to the Isle of Wight with plenty of company about in the form of schooners, boomies and other barges. Leaving Newport we went across to Poole where we loaded clay for Antwerp. Clearing Poole harbour at 6am on a Sunday morning, we were at Antwerp by 9pm on the following Tuesday evening. It was a really excellent passage for we were under way the whole time; perhaps not so remarkable considering that Antwerp lay on the West Schelde, a fast running river which always demanded a commanding breeze. Nevertheless, I hardly need say that we were dog tired and ready for a rest.

After discharging our clay we loaded silver sand for Aveling & Porter's at Strood. We worked our way down from Antwerp to the Rammekens, a little river above Flushing, and then onto Zeebrugge where we had to lay under the mole, sheltering from the strong south-westerly wind, in spite of the lack of room. We got a lead from the wind out to the Wandelaar light vessel and then on to the West Hinder. Although we had our topsail down the strong wind really drove us along, sending plenty of water and spray flying around, but we held our course until we sighted the North Foreland.

Standing down towards the Longnose buoy, we anchored for a tide, glad of the opportunity to get our wet clothes off and to have a square meal. We eventually fetched up at Strood on September 7th 1923, after a round voyage from the Medway to the Wight, then Wight to Poole, Poole to Antwerp and back to Strood all in six weeks.

Our next trip took us up to Erith to load loam for Newhaven. After loading we reached Dover and there we lay in company of many other barges. If a period of one's life was ever wasted it was during such times as these. Days and days went by, the first were spent cleaning up the vessel. Then it was fishing and reading and sleeping. Other times when we had plenty of room I even set up my own miniature rifle range, setting up the target on a bitt head. On one occasion, to make things a little harder, I set up a spent bullet as a target to fire at. After hitting it, a piece was nicked out of my ear by the ricochet coming back. That taught me a bit of a lesson and needless to say I packed in the rifle range. On one of my visits to the continent I had brought a violin and taught myself to play by ear, shut away in the after cabin. The mate stayed for'ard at such times, whether he loved or hated my playing, he never did say! For me then, it was anything to pass the time away, especially the long winter nights. Many other skippers and mates would go ashore when possible, to visit the pubs, but I never did drink.

At the first sign of some light airs we put out from Dover, leaving the other barges still there. In spite of some fog patches we finally made Newhaven. There we had to lower our gear down in order to go up the river. I took on two local men to give us a hand, but they were fishermen rather than bargemen, which still left us much to do for ourselves. We moved slowly up, punting with our setting booms. About halfway to our berth we discovered that we were rather too early, for the rudder suddenly caught the ground, breaking the casting on the rudder head. At Cliffe Bridge, Lewes we had to make fast and wait until the next tide before passing through. The barge had to be pulled through the span going up as well as when we dropped back on the ebb. In later years the *Shamrock* got stuck under the bridge when doing the same thing there. The fire brigade pumped her full of water to relieve the pressure on the bridge as she rose on the tide. Both barge and bridge survived, but the Shamrock had serious damage to her deck beams.

The *Shamrock* wedged under Cliffe Bridge, Lewes. (Newhaven Historical Society)

After discharging the loam at Phoenix Foundry, we returned down river to Newhaven and received orders to go over to Antwerp to load steel for Lewes, our freight being eight shillings per ton. We left Newhaven before a fresh nor'west wind, passing up the coast by Seaford and the Seven Sisters, and then on by Beachy Head to Dover where we put in for a night's rest. There we found still waiting, all the barges which had been previously becalmed with us. Early on the following morning we were again on our way out past the South Goodwin light vessel with a strong westerly wind behind us. That took us right over to the Ruytingen light vessel and so on to the Wandelaar. Off Zeebrugge we were hit by a heavy rebound of the swells from the mole, but we managed to fetch up in the entrance to the Schelde by dark and came to anchor to await the tide. We had to go over to Flushing the next morning to pick up the compulsory pilot, and with the wind falling rapidly away, we only just made Antwerp that day.

We loaded our cargo and went back down the river as far as the Rammekens, where we lay a few days before getting a fair wind which carried us across to Dover. As we turned away from the Kent coast to head down Channel we had light winds out of the west, but having reached a point a little to the west of Hastings the wind suddenly freshened and blew strongly with more south in it. We were forced to turn and start running back with the intention of coming to anchor under Dungeness. Having to gybe in a nasty swell off Dungeness, the chain pendant on the port vang carried away with a crash and went overside, followed by all the gear, the sails and the mainmast, which had broken in two and confirmed my earlier anxieties about its condition.

I set the mizzen quickly, while the gear overside acted as a sea anchor. After hoisting distress signals on the mizzen, the Dover harbour tug Lady Brassey came out and towed us back to harbour. We lay in Dover for some weeks waiting to have a new mast stepped and a general refit. *Knowles* was a stiff barge and was inclined to get on her head when running. While in Dover my mate left the barge and so I had to find another. Eventually, all was back together and we were ready to put to sea, taking advantage of a cold easterly wind which took us round to Newhaven. Again we poked our way up the narrow and tortuous channel with its awkward bridges, until we finally reached our berth in Lewes.

The Dover harbour tug
Lady Brassey.
(Photo: A Duncan,
Tony Farnham collection)

We did not accept any more freights for there as I considered it was not a safe river to navigate. So after coming back to the Thames light we traded between Ipswich and Rochester until the end of March, when my transfer to the *Scone* occurred. I had been in command of the *Knowles* for almost a year and in that time I had covered thousands of miles under sail, to the West Country and East Anglia, as well as over to the continent.

CHAPTER 8

Financial Arrangements

"... some skippers had a convenient lapse of memory when it came to paying their mate."

Most barges sailed by the share; that is a half share of the net freight for the crew. River barges under 45 net registered tons were exempt from paying Thames River Dues so there were no expenses to be set against the freight. For example, suppose a river freight was £9.0s.0d.[1] with no expenses. The owner would therefore claim £4.10s.0d. leaving the balance of £4.10s.0d. to be divided between the skipper and his mate. The mate would receive £1.10s.0d. and the skipper £3.0s.0d. Barges over 45 tons, which of course included all the coasters, had to pay dues, commissions on freights, towage fees, hired help and so on, while over 80 net registered tons were called upon to pay Light Dues as well. The rates for most of the outside work were laid down in the 'Blue Book', while those for the river were set out in the 'Pink Book'. These books were published jointly by the trade union and the barge owners.

The freight rate would of course vary according to the nature of the cargo, but £45 was typical for a 160 ton barge bound down to Ipswich. From that gross freight, various expenses had to be paid, such as the agent's fee in London, a commission, London dues on the vessel, dues at Ipswich and the cost of any towage or hired help of any kind. These expenses could quite easily add up to £5 leaving a net freight of £40. One half went to the men working the barge. That £20 was split up by the master taking two thirds, £13.7s.4d., and the mate one third, £6.13s.8d.

When it came to settling up, some barge owners were inclined to mis-judge the ability of their crews to do the sums! We once loaded a cargo of sugar from Ipswich to Wandsworth, secured at 4/3d per ton. I found out that the charterers were getting 4/9d per ton, only 3d less the Blue Book rate of 5/- per ton. I argued that we were entitled to half the freight at 4/9d per ton, and as a result ended up with £11.17s.6d. instead of £10.12s.6d. in my pocket. Bargemen may have not worn collars or ties, but we certainly knew what was due to us.

Even some skippers had a convenient lapse of memory when it came to paying their mate. My first experience of that was while serving on the *G.W.*, for although I was being paid my share of the Pink Book freight schedule, the actual freight rates had grown enormously due to the war. My share of that increased rate was not paid over to me. It was when that little fiddle of my skipper's was discovered by me that I had said goodbye to the *G.W.*

[1] *Throughout the text money is shown in Pounds, Shillings and old Pence. Shillings (s) were the equivalent of today's 5p (20 to the £), old Pence (d) around 20% of New Pence (12 to the Shilling). Sums less than a £1 were shown in Shillings and old Pence divided by an oblique, eg. 6/10d (c.34p), 16/6d (c.82p). A Guinea was £1.1s.0d., or £1.05 in decimal currency.*

Calculations of freight money and wages in 'Harry' Bagshaw's note book for a trip from Plymouth to Par with wheat. The freight was 130 tons @ 3/- which grossed £19.10s.0d. with expenses of £3.1s.7d. covering telegrams and stamps, pilotage and help. The £16.8s.5d. net freight money has been halved and a half guinea gratuity added, the mate's entitlement to a third appears to have been 'rounded down'. (Albert Bagshaw collection)

CHAPTER 9

Leaky
Scone

"A large sailing ship showing no lights at all was bearing right down on us."

My transfer to the *Scone* occurred in April 1924, when she was only five years old. Taking command of her meant that in just two years I had gone from the smallest barge in the owners fleet, to the largest.

Like the *Knowles*, the *Scone* was built by her owners, the London and Rochester Barge Company, having been launched at Frindsbury in 1919, with a Registered Tonnage of seventy-four and forty hundredths. She had a length of 88 feet, a beam of 21 feet and a depth of 7 feet 9 inches. The original fitting out had been done with old spars on account of scarcities caused by the war, including a boomie's mast which was certainly not to my liking. Memories of my experience with *Knowles'* mast were still fresh in my mind. I told the owners of my concern and that resulted in a new mast being ordered, which was eventually fitted some twelve months later.

Scone did not have an exceptionally good sheer, but was nevertheless a good sailing craft, which steered well. When I first boarded her she had been loaded with cement at Cliffe, and was waiting on the buoy at Grays ready to make the trip down to Torquay. The mate, Ipswich born Arthur Lambert, was already in her; he had been since she was new. He remained with me for six years before finally being compelled to give up the sea at the age of sixty-nine owing to ill health. The next mate stayed with me even longer for we were together no less than ten years. Our third hand was a young boy from Belvedere, Robert Stanifold, who was only fifteen years old, but was a capable lad, both about the barge and with the cooking.

Scone under way in the 1920's under 'Harry' Bagshaw. The spray from her bows has soaked the clew of her jib.
(Albert Bagshaw collection)

It didn't take me long to find what a good sailor *Scone* was, and when there was not too much swell I could trim her so that she would sail herself. Even with a moderate swell she only wanted three or four spokes of the wheel to keep her on course.

With a strong easterly wind blowing I brought her down to Northfleet in order to stow my gear aboard and take on our stores for the trip. When the easterly began to moderate we headed down river, rounded Lower Hope Point and went on into the Prince's Channel, as I decided it would be unwise to get to leeward round the North Foreland. With a nice north-east breeze and a smooth sea we made our way down Channel in good fashion, learning more of my new charge as we went. Passing St.Catherine's Point we shaped a course for St.Alban's Head.

In Weymouth Bay we hove to, waiting for the tide to take us round Portland Bill with a good weather shore. In order to keep inside the race we

43

A page from the Official Log Book for *Scone* covering the period July to October 1924. Although a large barge by the standards of the day, she still drew only 2 feet 6 inches fore and aft when light laden. It can be seen that her deepest recorded draft during this period was not for her freights in and around the Thames Estuary, but when she went down Channel to Torquay in November with just 12 inches freeboard. (Crown copyright, Public Record Office, Kew)

sailed close inshore, almost touching the rocks which swept down into the sea, for there was three fathoms of water close into the Bill. Away on our other hand the Race was troubled and turbulent water. We arrived at Torquay after a good first trip; so much better when compared to my last visit down Channel. After our discharging we received orders for Charlestown in Cornwall, where china clay was to be loaded for Queenborough.

The *Scone* was not as tight as the *Knowles* although she was certainly a much better craft to handle. We had to pump her out every day and the mate who had been in her since she was built told me that she had always been like it.

The diminutive Charlestown harbour was a 'wet dock', where vessels locked through on the high water and lay afloat to load. (Royal Institution of Cornwall)

Clay was loaded for mainland and continental destinations. Trucks would unload the china clay into chutes high on the east side of the harbour, which discharged directly into the holds of the waiting craft. The *Lady Rosebery* is nearly loaded, her crew were obviously aware that a camera was pointed at them. (National Maritime Museum, Greenwich, London)

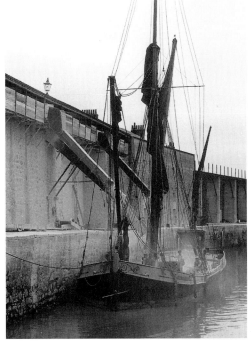

Apparently it had got a little worse since they hit a sand bar while in the Exeter river on one of her early runs down Channel. That was not to my liking at all and I made up my mind to locate the reason for the leakage and put an end to the continual pumping. After all, we were paid for getting her about, not keeping her afloat! After I had a number of repairs carried out in the early months of my stay in her, she proved to be a tight craft, and it was very rare that a claim for water damaged cargo was made against us.

Leaving Torquay, we rounded Berry Head in a nasty swell, whipped up by the wind which was still blowing from the east. We were soon passing Dartmouth harbour and then on to Start Point, Bolt Tail and so to Plymouth. Navigating sailing barges in light laden conditions in a seaway required an allowance of 6 to 8 points leeway. But when loaded, that allowance was reduced to about 3 points. A barge never fetched up at the point to which she was headed, but moved a bit like a crab across the water.

At Plymouth we put in for the night and carried on with our run to Charlestown the next morning. On our way down the coast we passed the picturesque harbours of Looe, Polperro and Fowey. Then we rounded Gribbin Head into Charlestown. Charlestown was a very small harbour lying in St.Austell Bay and was entered by lock gates between piers. Local boatmen known as hobblers came out to take vessels into the harbour. Once through the piers there was a sharp turn to starboard. After getting in without trouble we rigged up sheets to protect the decks before starting to load, otherwise the clay found its way everywhere. That was another of those places where the freight money was paid away to get in the cargo. It was not so bad as at Teignmouth, but nevertheless it called for hard work in rigging up the vang tackle on the forestay runner for the fore hold, the staysail halyard for the middle hold, and the topsail sheet for the after hold. After loading 170 tons of clay we cleared up and sailed, arriving at Queenborough about five days later; a round voyage from the Thames and back, all in three weeks.

Our next orders were for the Millwall Dock to load linseed for Ipswich. When we came alongside the ship however, she only delivered about 100 tons aboard, and so we had to go to Younghusband, Barnes & Co.'s Wharf at Rotherhithe to top up our hold with empty linseed oil drums which were also destined for Ipswich. On our return we carried linseed oil for various wharves, including Farmiloe's Wharf up under bridges at Nine Elms, for which we gained extra freight money, although that only paid the cost for towage.

Lower King & Queen Wharf operated by Younghusband, Barnes & Co. where empty barrels, which can be seen on the edge of the wharf and in a stack to the left, were loaded for Ipswich.
(Courtesy of Museum in Docklands, PLA collection)

After delivering the first four cargoes in *Scone*, I visited the owner's office to settle up our business expenses, where I found that the Director did not want to allow me the wages paid for my third hand. Of course, different firms paid different allowances, but while serving in the *Knowles* I had been allowed 8/- per week towards the wage paid out to the third hand. With the *Scone* it appeared at first a different matter. The Director turned round on me with the remark "Eight shillings! But look at the money you are earning out of my property." That was one of the first occasions that I experienced the short sightedness of a fairly rich man. My answer was to the effect that if I personally was going to be 8/- per week worse off for handling a larger barge then I would rather take over a smaller one again. Nothing more was said at the time, but obviously my point had been well made as I did receive the allowance for the third hand.

Later in the year we had put into Ipswich with a cargo of wheat. Other bargemen sounded me out about joining the Union in order to get the thankless task of cargo working abolished. Remembering my experience of that as a young lad I agreed wholeheartedly and for years after that I regularly paid my union contributions due every three months. True enough, the Union was as good as its word and cargo handling went, but nevertheless, on the only occasion on which I required help from the Union over the question of a freight rate, my enquiry for assistance fell on deaf ears in spite of my regular payments. It was the same old result, one had to stick up for oneself, for the Union cared little for the individual.

Soon after the Ipswich trip we loaded a cargo of timber in the Surrey Commercial Dock for Ramsgate. After successfully discharging the 35 standards

The continuation of Scone's Official Log Book through to the end of 1924. Again she has just 12 inches freeboard for the passage from Charlestown to Rochester, laden with china clay.
(Crown copyright, Public Record Office, Kew)

Official Log of the *Scone* (continued). 1924.

Date and Place of Occurrence.		Forward	Aft	Dovi	Date and Place of Occurrence.	Entry.
Nov 13	Torquay	2.6	2.6	5.0	5.0	Charlestown	14 Nov
Nov 17	Charlestown	6.6	7.0	1.0	1.0	Roch:	25 Nov
Nov 24	Rochester	2.6	2.6	5.0	5.0	Tilbury (Po) London	Dec 1
Dec 8	Tilbury	5.0	6.0	2.0	2.0	Ipswich	Dec 6
Dec 12	Ipswich	2.6	2.6	5.0	5.0	London	Dec 15
Dec 19	London	5.0	6.0	2.0	2.0	Ipswich	Dec 21
Dec 27	Ipswich	2.6	2.6	5.0	5.0	London	Dec 29
						A. B. 2 jshaw	

[1] *The floor of the hold was known as the 'ceiling', occasionally spelt as 'sealing', though opinions differ over the origins of either term.*

[2] *Baulks of timber anchored to the riverbed which supported the barge when the tide ebbed, allowing access to the underside of the hull.*

Kathleen Holden, pictured aged 18, when 'employed in general service'.
(Albert Bagshaw collection)

of timber we returned to London light. There was a strong easterly wind blowing when we nosed our way out of Ramsgate harbour, and we sustained several hard knocks from the sea before we rounded the North Foreland. The wind was in our favour, of course, for the run up to Gravesend, but here the wind was far too fresh to run the risk of bringing up to our anchor on a lee shore, so I carried on up to a safe piece of ground off Grays. As we stowed away our sails I sensed that something was wrong with the vessel, and it was not long before we realised that she was listing.

On taking off the hatches, sure enough, there was water over the ceiling[1] in the hold. After reporting by telephone to the office we received instructions to proceed to the yard at Strood. I purposely left the water in her for that trip, and it made her lay her ears back as we went down Sea Reach. After arriving at Strood, the Foreman Shipwright who had originally built the barge refused to believe that she was not tight. "Well, as far as I am concerned the water is in the hold, no doubt it will come out the same way as it got in." I told him. The blocks[2] had been cleared for us by the time we arrived so on we went. As soon as the water left her we went under and there was our proof; the water already running out of the bottom planking, through the butt joints in the timbers. It was clearly the cause of the water in her, which the mate had spent long hours pumping during the previous five years. The butts were caulked, and covered with galvanised steel plates and it proved to be a total cure.

From Strood we sailed down to Chatham where we loaded 150 tons of flour for Plymouth. After loading we went down to Sheerness where we were forced to lay idle for a week owing to lack of favourable winds. It gave me the chance for some time ashore, much of which was spent in the company of Kathleen Holden, with whom I was walking out. It was a chance to plan for our wedding which we decided should take place in the new year. It was during that week while ashore that I learned that we were fixed to load clay out of Teignmouth. With the previous trip there still fresh in my mind, I was determined not to experience a repeat performance of the *Knowles'* cargo problems. With promptitude I put my case forward, saying that if the barge must load there then she would have to load under another master. The need of a decision never arose however, for weavils were discovered in our cargo on arrival at Plymouth, and we were ordered to return with it to Chatham.

We cast off from Devonport Dockyard at 1pm on the Wednesday and arrived in the Medway at Long Reach at 8pm on the Friday, a run of 55 hours, during which we made Beachy Head from Start Point in 24 hours. On other occasions I made that run in 20 and 22 hours, which was really good sailing.

On the run up from Plymouth with our cargo of unwanted flour we received quite a nasty shock. I had installed a large square sail, set from a yard hung under the lower mainmast cap. With the square sail set and a boom in the mainsail we ran with a strong south westerly wind at some speed. We were thrusting through a nasty sea when I went below for a cup of tea. Suddenly the mate sung out to me to come on deck fast. Dropping everything I climbed the ladder with apprehension. Leaving the oil lit cabin, my eyes had to get accustomed to the pitch black night. As I reached his side I saw the cause of his panic. A large sailing ship showing no lights at all was bearing right down on us. We let go the square sail forward to bring the barge on to the wind under the other vessel's approaching stem. We managed to luff clear of our mysterious adversary, which passed us just a barge width to leeward and went away into the blackness of the night. I could not help but compare in my mind that event with the stories of the Flying Dutchman which appeared and disappeared before the gaze of ancient seafarers.

As could be expected, there was a good deal of argument over the condition of our flour cargo when we arrived back at Chatham, but I never did hear the final result. For our part, it was enough for me that our freight was paid up in full. For the remainder of 1924 we traded to the nearer ports of Ramsgate and Ipswich. At the beginning of December we got regular work between Ipswich and the Thames and Medway, which was to continue for the whole of 1925.

Barges loading ex ship in Ipswich Dock. They would lay inside, between ship and quay, as well as outside the discharging steamer. (Richard Smith collection)

CHAPTER 10

Duty
Calls

"... we bashed our way down as far as Beachy Head, then we were forced by the bad weather to run back to Dungeness."

With the New Year of 1925 five days old, we loaded 189 tons of linseed in the Victoria Docks, London for Ipswich. Sailing late on the 7th, we were down the coast and in the Orwell next day and locked in on the 9th, completing unloading by the 12th.

The loading and discharging time allowed in the Union Blue Book was seven days. Although that book laid down the rates, it was of little practical value in those days. It gave 5/- per ton to lift sugar, yet I was only paid 4/3d, which showed the ineffectiveness of the Unions. Freights to Yarmouth were also cut below the Blue Book rates, and on top of this there was stiff competition from the small Dutch craft which were always willing to do the work offered at almost any price. The barge had to take what was offered to earn anything at all, especially as bargemen were not eligible for any dole money. In spite of having to pay contributions, the fact that we sailed by the share denied us the benefit.

The round trip from London to Ipswich was rarely made under fourteen days, the loaded outward run taking about eight. It was the custom for the barge to be under the dock merchant's big elevator at 6am with the hatches off ready for the days work. More often than not we would be empty before noon ready to make our return run light ship.

On the 19th January 1925, we loaded about 148 tons of cotton seed in the Millwall Dock. Not only had we to trim the freight, we also had to erect a large stack on the hatches as a deck cargo. That was really hard work and took about four hours. Dust from that type of cargo would get into eyes and throats making them very sore. It gave me a kind of hay fever with running eyes and nose, making it very unpleasant. Loading was completed late on the 23rd and after sailing during the hours of darkness we arrived at Ipswich on the 25th. We were empty by the 27th and we set out on the return passage, bound for Millwall Dock again where we were to load ground nuts, or as some people call them, monkey nuts. We started on the 2nd February, finished the following day and were at Ipswich again on the 5th. It was not until the 13th that we had completed discharge and so four days demurrage[1] was incurred. After that it was back to the Millwall Dock where a further cargo of cotton seed awaited us. That time our destination was Strood where we arrived on the 19th. From Strood we returned to Cook's Wharf, Millwall where we took on more cotton seed, that time for Chatham. Loading was completed by the 10th March and our cargo had been worked out by the 17th. As you can imagine the hay fever effect was pretty bad by then.

By way of a change, our next run took us to Torquay. We commenced loading our cargo at Cliffe on 23rd March 1925, a date unlikely to be forgotten for it was the day on which I took time off to get married. We never did get a

[1] *A compensatory payment for vessels unable to discharge within an agreed period due to delays attributable to the receiver of the cargo.*

'Harry' Bagshaw and Kathleen
Holden on their wedding day.
(Albert Bagshaw collection)

honeymoon, however, for the loading was finished by the 25th March and we left that day for Torquay. Our hatches were off on the 30th and we were empty on April Fool's Day. We earned 11/- per ton for that trip, our cargo being 170 tons of cement, yielding £98.10s.0d. against which our expenses came to £10.2s.6d.

From Torquay we sailed light to Charlestown where we took aboard washed clay for papermaking at Reeds of Aylesford. Arriving at the Cornish port on the 3rd April, we finished loading on the 6th and were at Strood by the 13th and discharging our cargo up the Medway at Aylesford over the 15th and 16th. That time our cargo weighed 170 tons 16 cwt, and earned a freight of £85.8s.0d. at 10/- per ton, against which there were rather higher expenses of £16.10s.8d. to be set.

Continuing to cover long distances in relatively short time, we left the Medway for Greenwich buoys to load rice bran for Strood. Arriving on the 24th, we had loaded by the 27th and were back in the Medway on the 28th. By 1st May we were ready for our next run which was a cargo of asphalt. We had finished loading by the 11th at Bow Creek for Northam. We arrived there five days out from London after working our way up Southampton Water and lowering down our gear in order to pass under Northam Bridge. After discharging our orders were to go to Bridport, a little harbour known as West Bay about 15 miles west of Portland Bill. With light winds we sailed out of Northam and came down to Yarmouth, Isle of Wight, on one tide, where we anchored to wait for the next. The whole operation sounds rather absurdly simple, yet it was not just a question of letting go the moorings and moving away. It was hard going, for we had to lower the gear; then once clear of Northam Bridge we had to heave up and re-rig before preparing to make our way out through the channel, which was not only narrow but filled by large numbers of yachts. To get the barge into a good position to set off, full sail was set to try and get her going with decent steerage way. Finally we made headway and as the water gurgled under the quarters we had a pleasant run down Southampton Water.

Next morning we got under way with our course shaped for Bridport. The wind being light, we went out through the inside of the channel by Hurst Point, steering between the shore and the shingle bank. In fact, so long as the barge's head was kept towards the shore, it was quite safe to drive out by that way until the Christchurch Ledge buoy was reached. There we were fortunate in picking up a nice north-west wind which took us in fine style to the race round St.Albans Head where pieces of the wrecked steamer Treveal could be seen lying under the headland. Sailing across Weymouth Bay, we passed close in to Portland Bill and so on to West Bay where we had to dodge about waiting for water, as the harbour dried out at low tide. Still with the light north-west wind,

The narrow entrance to West Bay harbour, Bridport, was vulnerable to silting up and leaving craft trapped within.
(Tony Farnham collection)

under just our foresail and mizzen, *Scone* kept fetching about backwards and forwards waiting for the flood.

As soon as the tide had made sufficiently we stood into West Bay and by the 26th May we had finished loading. Then followed several days of enforced idleness, for the strong winds and heavy seas during this time had been doing their best to keep us land-locked. They had pushed huge heaps of shingle into the narrow harbour entrance, so much that the mouth had become blocked. It was some few days later, after those furious seas had calmed down, that the accumulation that had built up was dredged clear. It was the 8th June before we arrived back at Blackwall on the Thames. We finished discharging a couple of days later, after having been the best part of a month on the round voyage.

Then followed some short passages, beginning with a run from Bellamy's Wharf to the Medway with pollards, then back empty to load cotton seed at Cook's Wharf for Strood and next a timber cargo from the Surrey Dock to Gillingham. When we finished unloading that cargo on the 28th July we went onto the yard for repairs. In addition to all the running and standing gear being renewed, we also had a new mast. From then on it became the practice for the barge to be equipped with a new set of sails every third year, and all new running ropes every second year, while all standing wires, including backstays and topmast stays, were renewed every fourth year. Other items such as vang falls, mainsheet, and the davit falls were renewed each year, as were the hatch cloths. Just before the winter closed in all mooring ropes would receive attention and were made good where necessary. *Scone* was kept well found; it was seldom that we broke any gear through lack of attention or cost of replacement.

The new mast which was fitted at that time measured fifty feet overall in length and was 13 inches thick. From heel to hounds was 39'6", so the masthead doubling[1] measured 10'6". It was indeed a lovely spar. That much needed refit lasted for three weeks. When we left the ways on August 25th we

[1] *Where the top of the mainmast and the lower part of the topmast overlap.*

were ordered up to three loading places in London; the East India Dock, the Millwall Dock, and then the Dock Head. We were finished loading by August 31st and sailed for Plymouth where we arrived on September 12th.

There was a cargo of wheat for us which we lifted from Plymouth for Par, another little harbour in St.Austell Bay which dried out at low water and was exposed to south-easterly winds. Our arrival at Par was on the 19th and our discharge completed by the 22nd, when we loaded clay and stone for Queenborough. Par was left astern on the 25th and after going up into Plymouth for one night, we arrived at Queenborough on the 29th.

It was on that voyage that a fresh north-west wind gave me one of the best runs I ever had; from Start Point to Beachy Head, a distance of 150 miles in just twenty hours. We also had our worries on the trip however, for just after change of watch when we were abeam of Beachy Head the mate sang out for me to come up from below. On reaching the deck I saw that our boat was hanging precariously and being dragged along through the water. This was the result of a sea that had broken aboard us and had filled our boat which was slung by the falls in the davits. The extra weight of the water had caused the after davit to bend dropping the stern into the water. Quickly I pulled out my knife and as the barge lifted to a sea I cut the after fall which allowed the boat to hang from the for'ard davit and empty itself. The other fall and guys were also cut and the boat, then on an even keel in the water, fell astern held in check by the painter. We then stood into smoother water off Eastbourne where we hauled the boat aboard. Luckily it was undamaged.

After we had discharged at Queenborough on the 30th September, only four weeks since we had first left London for the West Country ports, our next orders were to proceed up river to the Surrey Commercial Dock to load barley for Ipswich. We anchored off the Ship and Lobster pub at Gravesend for a tide

The Ship and Lobster public house at Denton below Gravesend, one of a number of Thamesside hostelries favoured by bargemen. (Kent County Council, Gravesend Library)

for my wife to come aboard, for at last she had condescended to come up to London with me. During our stay in London we went to the theatre and saw the musical, Showboat. In the interval Kathleen thought it very amusing to see me in my high neck dark blue seaman's jersey eating boxed chocolates, while others around us were dressed up in their best bibs and tuckers.

Back on board the loading was completed by October 9th and we made our way down river. My wife turned to me with a questioning look as we passed Gravesend, for that was where she had expected to be put ashore. "I'm sorry love, you are Shanghaied. You are coming to Ipswich with us." "But I have no change of clothes." was her answer and her main concern. "Don't worry about that" I said "for we can soon buy some when we get there."

With the weather set fair we had a lovely passage down, much to Kathleen's delight, and we reached Ipswich on the 14th. As soon as the barge was berthed and could be left, I went shopping. "Can I help you, Sir" said the shop assistant. "I would like a full change of clothing for my wife please, as she has fallen overboard." I replied. "What size is the lady?" she asked. Fortunately for me the young lady looked about the same size as my wife. "About your size." I answered. I thought later, wouldn't I have been in a dilemma if she hadn't been. When I handed over the new clothes to my wife it was quite a joke with us.

We returned light to Cliffe where my wife left us, having enjoyed those few days aboard. That was the first of many trips she was to make, along with our children in later years.

We finished loading cement for Torquay on October 22nd, but we were not able to get there until November 13th, for we had a very dirty passage. We had to put into Newhaven wind bound. It was a very wild night and I had the greatest difficulty in keeping the barge under control. We had the misfortune to damage a number of yachts that were moored there. Torquay was reached eventually and our cargo discharge completed by the 18th November. Leaving Torquay we shaped a course for Par where we arrived two days later to load clay again. That cargo was destined for Chiswick, up beyond many of the bridges spanning the Thames. I had requested 170 tons to be loaded but the cargo turned out to be in bags. The over weight, for which I had not made any allowance, made the total cargo come to 179 tons and 15 cwts. That made the barge lie very deep in the water for a long passage in the unpredictable days of winter.

The harbour at Newhaven provided refuge for *Scone* on her way down Channel to Torquay. (Tony Farnham collection)

We sailed from Par on November 27th with a strong north-east wind blowing right off the shore. After weighing up all the pros and cons I decided we could carry on right up to Plymouth. We had a good sail with the wind holding until we were off Plymouth, when it veered round a little to the north, making it fair for Start Point. We put up our lights and kept on going. By daylight the following morning we were close in under Portland Bill, having made the run across the bay in fine style. As we passed round the Bill and inside the Shambles lightship the sky began to get very black from off the land. Bearing in mind my earlier experience of squalls, we ran down the topsail and foresail and took the mizzen off as well as a bit of the mainsail. In spite of our reduced sail area we still maintained a good course up Channel and sure enough the squall caught up with us and brought heavy snow with it. Needless to say I was very glad that we had snugged our barge down in good time. By the time we had made St.Albans Head the squall had passed over so we made full sail again and bowled along to Anvil Point with the wind blowing strongly from the north-east. We held up well to windward and found ourselves at the Christchurch Ledge buoy before passing inside the Shingle Bank.

Rounding Hurst Point we just saved our tide and came to anchor at the Jack-in-the-Basket, a little above Lymington Creek, where we at last had a good night's sleep. The following morning saw us on the way to Stokes Bay where we spent the next night, and with the tide just right, the next morning we were under way again. With the wind blowing strongly from the north-west we passed out through the Forts and up through the Looe Channel. But that favourable wind was not with us for long before it backed to the south-west and began to give us a pounding.

Even with the head of the topsail down and just half of the mainsail, we were taking some heavy seas aboard. Steering a little to the south of east we passed the Royal Sovereign light vessel after passing Beachy Head and shaped a course up for Dungeness. The seas were still running high but we had good gear and the worst was probably behind us.

There were just the three of us, two men and a fifteen year old boy to do battle with the elements and bring our barge and cargo safely into port. In winter it was more than ever essential that everything was stowed up right. On the *Scone* the jib was stowed on the bowsprit and of course the topsail stowed up aloft. Invariably there were 30 fathoms of chain out when we anchored which all had to be hauled in by hand. Nevertheless, in later years when freight rates fell to very low levels even this meagre crew of three was depleted by our inability to afford the wages of the boy.

After passing Dungeness the wind was still inclined to back to the south'ard and so we kept well off to the Varne light vessel. The South Foreland light was then sighted, but it was inadvisable to approach Dover too closely as the rebound from the breakwater would have made the sea very bad. In spite of keeping well clear, the seas began to sweep over our barge from end to end as we came abeam of Dover. So bad was it that we had to keep a close watch on our side lights to make sure they were not doused, but still burned brightly. That was some job in that type of weather. Happily the light for the compass posed less trouble as it was lit by the cabin lamp. With the wind backed around strongly to the south-east we sighted the South Goodwin light vessel. In order to fetch through the Downs we had to heave in our vang and main sheet. There was no respite for us that night, for the wind was blowing strongly into the land and made an anchorage impossible.

Eventually we rounded the North Foreland and by the time Margate Pier light had opened out the wind had veered to the east, but was blowing as strongly as ever. I had to try and go very easy, for I wanted daylight to come in order that I might find my way up the Gore Channel. However I decided not to risk it and stood away to the North-East Spit buoy, making a good run up the Princes Channel to Gravesend where my wife, of whom I had seen very little, was waiting for me.

We were up under the bridges to Chiswick by December 3rd. When I visited our firm's London office I was asked where my barge was. When I said that she was discharging at Chiswick and had quoted the details, I was told that during the time we were beating our way up Channel, coasting barges such as ours had been unable to leave Sheerness for the short run up to London! Our unloading was completed on December 8th, allowing us time for one more trip to Ipswich before the year of 1925 finally drew to its close. In all, the twelve months had seen us carry twenty cargoes.

Four days of the New Year had passed when we arrived at Ipswich with cotton seed, loaded in the last days of the old year at Millwall Dock in London. It was the usual run out of the Thames Estuary, a run made very many times, but in many different conditions. We might be favoured by fair winds or frustrated by a breeze on our head. At other times the winds would be light, or fog might roll in, when anchor drill would be the order coming down the river.

By January 14th we were back in the King George V Dock loading maize for Strood where we arrived on the 18th. Then it was back to the Millwall Dock to load for Strood again, arriving there on the 28th. That all sounds rather matter of fact but, as with many other places, the navigation of the Medway could be tricky. It was often an anxious time bringing a barge up Chatham Reach, Limehouse Reach and Bridge Reach, past the numerous destroyers and other naval craft ever present at the various buoys. Coming down the river presented many similar problems, such as rounding Gas House Point where a strong wind was needed, but when it came it was often not very steady in strength or direction.

After discharging at Strood we returned to the Millwall Dock to load maize for Dover. Owing to the congestion of lighters it was not at all easy to move about in the Millwall Dock, especially if the wind did not serve. With a strong wind astern it was always difficult to check[1] the barge at the dock entrance when going in, although once locked through the wind was very useful in making a run up the dock. When going to the grain silos however, we had a bridge to contend with and this meant boat and line work, constantly heaving or checking the vessel. Of course at times we secured a tow, which entailed the payment of a few shillings, but which was very well worth it. All those problems were often encountered once again when leaving the dock. Sometimes we could sail straight down the dock to the entrance, and after the tug towed us out into the river we could get straight away. Other times we would have to come to anchor. With an easterly wind blowing and the tide on the top of the flood we could leave the dock and perhaps reach the bottom of Long Reach or even Grays by low water. Even better, when the wind blew out of the west, the turn of the tide at next low water would see us as far as the Lower Hope or even down to the Mid Blyth Buoy.

In the Millwall Dock 750 quarters of maize filled our holds and on the 10th February we set sail for Dover where we arrived on the morning of the 13th. By the evening of the 16th we had started back empty for the London River, receiving orders to proceed to the Free Trade Wharf where we commenced loading sunflower meal for Strood. We were finished by February 22nd and reached Strood the following day..

For years I had regularly paid my Union subscriptions without protest, and for my part without requesting any help, advice or support. That cargo for Strood consisted of no less than 2,371 bags of sunflower seed which filled the holds and in addition was stacked thirteen bags high on deck, which made

[1] *To take the 'way' off, or slow down.*

55

sailing very difficult indeed. The freight rate for sunflower meal did not appear in the laid down scales which appeared in the Pink Book, so that when it came to the payment for its carriage, I was forced to accept the rate for the carriage of cement, which was far heavier and demanded far less cubic space. There was just no comparison. It was therefore with a feeling of justification that I approached the local Union representative for help and advice. He showed no interest in the matter whatsoever, and did not attempt to make any enquiries on my behalf. That showed me the real priorities of the Union officials; the invidious principle that an individual's problems were of no relevance or interest and that the manipulation of the masses to the Union's political ends was their only priority. Needless to say, that finished me so far as Trade Unions were concerned.

Our next freight was linseed, and we carried 3,191 bags weighing 184 tons and 10 cwts to Ipswich, arriving there on March 6th and finishing our discharge on March 9th. The trip out of London round the Essex coast down to Ipswich followed a general pattern. Usually it meant a careful passage down the crowded reaches of the Thames until it opened up towards the sea at Sea Reach. Then we would sail on past Southend and the measured mile off Shoeburyness before making for the Blacktail Spit buoy. Thereafter it was necessary to pick out the line of buoys which led to the Maplin lighthouse, built on stilts like the Chapman and Gunfleet lights. We used to pass inside the light on the flood and if on the ebb, or with the wind in the east, we used to steer to the south'ard of it.

Once passed the light our course was changed to fetch up inside the Swin Channel and clear of the Middle as it was called. From the Swin Middle lightship we had to head for the Swin Spitway buoy. The Spitway was a channel through the sand from the Swin to the Wallet. From the Swin Spitway buoy it depended upon whether the tide was flood or ebb, the height of the tide, and of course the weather as to which course to take. Perhaps through the sands,

Many sailing barges loaded ex ship or alongside at Ipswich, but *Scone* often departed light. (Tony Farnham collection)

or fetching up to our anchor to await suitable conditions, or perhaps even to go outside and around the Gunfleet Sand and its lighthouse. At odd times barges would have to go round the Gunfleet, then the West Rocks and then down to the Cork light vessel, but this was usually when they were bound Great Yarmouth way. Most barges however, made a safe passage over the Spitway and then proceeded to a fair offing from Walton Pier and then onwards until the Naze was abeam before entering the Medusa Channel and setting a northerly course for Harwich harbour.

On entering the harbour, the town of Harwich was away to port, while on the starboard side lay Felixstowe. The ground was flat on the Harwich side and the entrance Channel was not altogether wide. Once inside, the mouths of two rivers came into view; the Stour which led up to Mistley, and the Orwell which led up to Ipswich and was also very commonly known as the Ipswich River. It was a particularly bad river to sail up with a north-west wind or in light airs as there was very little run in the tide. The first few miles up past Pin Mill were not too bad, but above Cat House Point and Redgate Hard things really became difficult as the Channel was quite narrow. Once over Redgate Hard it was not possible to turn the barge to windward without a very great risk of going ashore. It was necessary therefore, to fill the sails well with the wind and then shoot up to windward with the way that was still on the barge, then let her head fall away again to refill the sails and get a bit more way on to repeat the exercise. Once through this bit of the river the remainder of the way up to the docks was not too bad. Often we traded up into St.Peter's Dock which actually lies across the original channel of the river which runs through to Stowmarket. A small motorboat occasionally gave us a tow, for I had a fair enough working agreement which covered 'tow or sail' at the same rate.

After discharging on the 9th March we returned to London where we loaded potash at the Free Trade Wharf, Wapping for Strood. Loading took place from a ship alongside the wharf and we were in the Medway on the 19th. The following day we were once again empty and returning to Bellamy's Wharf, Rotherhithe, where loose linseed was loaded for Strood. Loading was finished by the 27th and Strood reached on the 30th. We were empty by the first day of April and went over to Gillingham to load potatoes, straight from a ship, for Southend. That was a cargo which demanded more than the usual attention for it was

Barges lay discharging at the Corporation Jetty, Southend-on-Sea. (Reproduced by courtesy of the Essex Record Office, Southend Branch)

made up with many different kinds; Edwards, Great Scot, Majestics, Arran Banners and so on, all of which had to be kept separate. With a bit of planning we succeeded in that regard and on the 7th we set out across the Thames Estuary. We arrived on the following day and the receiver of the cargo came aboard and asked "Could I by any chance have the Majestics out first, and then the Great Scots?" He went on to explain that they were all for various merchants and apparently he imagined that could give rise to a lot of trouble and extra work for us. There was little doubt of his surprise when I told him he could have his potatoes in the order he wanted.

From Southend we sailed up the Thames to Tilbury Dock where we loaded a cargo of rice bran. We sailed on the 19th April and arrived at Strood, our destination, on the following day. After discharge we made our way over to our company's barge yard where we went on the ways for a refit.

By the time we were ready for sea again the General Strike had commenced and we had orders to proceed to Antwerp light. We had only reached Margate Roads when further instructions were received verbally from a runner who came out to us, that we should return to load linseed in the Victoria Docks for Ipswich. On the 20th May we had finished loading 188 tons and on the 25th we were at Ipswich where we finished discharging on the last day of the month. It was back to Tilbury Dock that time where a cargo of rice meal awaited us. We loaded on the 3rd June and sailed for the Medway, where our cargo was discharged at Strood by the 9th, a total of 177 tons.

A really big freight, 200 tons of coal from a steamer at Woolwich Buoys, next filled our barge, bound for the drying harbour at Margate Gas Works. We arrived there safely on June 13th, started discharging straight away, and were sweeping our holds by the 15th.

After all those estuarial freights our next voyage was by way of a change a long one. After leaving Margate we went to the King George V Dock in London to load oats for Plymouth. Our cargo was 893 quarters and the freight rate was 2/- per quarter. We arrived in the dock on the 16th June, and finished loading by the 19th. There followed an uneventful voyage with fine weather and variable winds. In fact, we made a non-stop run from the Downs to Plymouth. The wind had been generally south-westerly and light during the day, and from the north-east during the night. That was too good to miss. We headed straight outside everything down Channel arriving at Plymouth by the 26th. We had been under way for five days and four nights and towards the end of the voyage I found the need to bath my eyes with cold water to keep myself awake. This was not the first, or by any means the only time for such measures. Needless to say by the time Plymouth was reached I was tired and only wanted to sleep. Our first son was born on the 29th June, and my wife never quite forgave me for not coming home to see him.

After discharging the oats by the 30th we had orders to proceed to Pentewan, which was a little harbour in the St.Austell Bay area. Standing into the bay first came Dodman Point to the west, followed by the harbour of Mevagissey, then the hole in the shore to Pentewan itself with Black Head and Cribbin Head, with its red and white striped light tower, facing one another across the Bay. Charlestown and Par harbours made up the centre. The artificial hills of unwanted clay could be clearly seen from the sea, especially those to the rear of Par. The Pentewan entrance was a very narrow one, but once beyond the lock gate inside, one could feel quite safe for a change. We arrived there on July 1st and were loaded and on our way again to Ipswich the following day. The passage took less than a week, despite the misfortune of a rolling vang which broke on our way. It happened when we were about half way between Portland Bill and St. Albans Head. We needed our rolling vangs rigged to hold the sprit steady working in the channel swells, and the

The picturesque Cornish port of Pentewan around 1926, with a Thames sailing barge and another vessel in the central basin. Although much of the harbour still exists, it has been cut off from the sea by a build up of sand for about forty years.
(Photo: Dr D A Lubbock, Dr M J T Lewis collection)

breakage could have meant serious trouble for us. We quickly stowed the mainsail and topsail and then hove in tight both main vangs, pulling the sprit to the centre of the barge. The mate was a little too old for the job, so I climbed up one of the vangs to the sprit end with a light heaving line slung over my shoulder. It was some job to get over the eight inch mainsail head rope and sit on the sprit end, which was easily 50 feet above the deck.

By a stroke of luck it was the port rolling vang that had broken, for the starboard vang was placed under the head rope and would have been much more difficult to get at. Holding myself on with both of my legs wrapped tightly around each main vang to free my hands, I took out the staple and removed the wire eye of the rolling vang from over the sprit end. I then sent it down on the line and hoisted up a spare one. I had to use a good deal of strength to avoid being tossed into the sea or onto the deck from my precarious perch, as the barge was leaping around in a heavy swell. Eventually I secured the replacement and so made my descent down the wire and over the blocks to the deck again, my legs trembling with exertion, and probably a bit of fear. This was not untypical of the sort of jobs which came the way of those who served on barges.

We arrived at Ipswich and while the barge was discharging I was off home by train to Gravesend to see my wife and our 10 day old son for the first time. The journey home was not seen by me, for the opportunity to catch up on some of my sleepless nights was not to be missed. After hellos and goodbyes, I was back at the barge on the 12th for our next freight which was to be Ipswich again, from Tilbury Dock. We were there and loaded by the 15th and back at Ipswich on the 17th. Once empty we had orders to proceed to Strood Oil Mills to load oil cake for West Bay, Bridport. A strong westerly wind was blowing when we left Ipswich and it was very difficult to turn up the Wallet. It was very seldom that I was forced to give up, but on that occasion we had to go into the Colne for shelter. Finally we arrived at Strood on the 25th and by the 27th we had loaded 181 tons. We made another smart passage and were down in Bridport in five days, arriving there on the 2nd August.

From Bridport we went to Portland to load stone. Stone was a new cargo for us. It was a bad cargo to carry as some of the lumps weighed anything between six and ten tons, although it was of course cleaner than clay. The charterer specified that freight would be paid to us on 150 tons after trimming. But after we had trimmed her with extra stone to avoid the risk of the large pieces shifting, we had nearer 170 tons aboard, extra weight that we were compelled to carry for the safety of our craft, but for which we were not to be paid. The last of the blocks was aboard on August 9th and we made a good passage to North Woolwich. There we had to lower the gear to go up through the bridges to the Portland Stone Company's wharf at Pimlico where we arrived on the 14th.

Lowering our gear was quite a task. The first job was to get the bowsprit inboard. On the *Scone* we had a standing jib on the bowsprit and the stay was fastened to the mast head by a block, while on the bowsprit there was a bottle screw to tighten it. This had proved a very secure form of rigging and by then I had gear aloft that I was confident should have stood any weather that we were inclined to sail in. So the jib had to be unbent and the bowsprit, which was a spar 28 feet in length, unshipped. It was held in the bitts by a steel bolt, so we had to use the heel rope tackle of the topmast and the long topmast stay to unship it. The long topmast stay was run over the port bow, so that the spar could be swung around to starboard, and then wound round the windlass. After putting the dogs on the anchor chain and slacking off, the heel rope tackle was put on the stem head and onto the heel of the spar. An eyebolt was fixed for that purpose, and by balancing the weight, the heel bolt holding the bowsprit would come out quite easily. Then the tackle would be slowly slackened off, while the topmast stay would be hove on, pulling the bowsprit inboard. The gear would be taken off, coiled up and stowed for'ard of the windlass. The staysail halyard would then be used to lift the bowsprit off the fore hatch and along the starboard deck out of the way.

The topmast was fairly easy to strike. The wire heel rope tack would be run over the winch on the mast case to lift the weight of the spar off the lower mainmast cap; then the boy would be sent aloft to take out the fid, enabling us to lower away. We would also have to ship our stem blocks to lower down, for our gear was normally held up by a large bottle screw. We had to do much of that procedure in reverse to clear our hatchways for unloading once alongside our berth. And then of course lower the gear again for the return run down through the bridges. Working above bridges for a large coaster with heavy gear was really hard work. All that for a cargo which earned us just £53.2s.0d. after expenses.

After discharging that nineteenth cargo of the year at Pimlico we went to the Victoria Dock where we loaded for Strood, finishing discharge on the 25th August. The next voyage was a long one, for we loaded on the 1st September for Newport in the Isle of Wight, but did not get down there until the 20th owing to strong winds. After lying windbound off Deal for a few days we decided, despite the weather, to try our hand in getting underway. One couldn't live swinging at the end of an anchor. Having made the effort, we bashed our way down as far as Beachy Head, but then we were forced to run back to Dungeness, and remained there until greeted by a favourable wind. Newport was on the river Medina which had its entrance at Cowes. There was not a lot of room to sail a large barge up that river, nor was there much water under her when above the cement works. At low tide the river dried out.

After unloading we were to go to Portland for another cargo of stone. With the wind coming from the south we tried to keep to windward, but after clearing the Needles and the Shingle Bank it fell light and we drove down passed St.Albans Head close inshore right under the cliffs. As darkness fell we were still very close to the land. The water was too deep for us to come to anchor, we could only keep the head of the barge off the shore by rowing with a large oar. Concerned as I was about our situation, we made ready a setting boom to fend us off, if it became necessary. Fortune smiled at us that trip and we managed to round St.Albans Head, where we entered the tide race. The stillness was broken as a light breeze picked up from the west and we began to gather way, making Portland by daylight.

Our return to Pimlico took nine days and we were emptied in a further two. Our next trip was to Ipswich with linseed again. We loaded in the Victoria Dock and were completed by the 9th, enabling us to reach Ipswich on the 11th. We set out on the return passage on the 19th with a light ship and orders to load coke in the Victoria Dock for Queenborough. That was an easy freight really, for loading was finished on the 24th October and our holds were free by the 26th.

From the Medway we were ordered to proceed light to Dunkirk. Owing to unfavourable winds we did not get away from Queenborough for several days, yet we made the French port on November 8th. We loaded gluten meal for Rochester where we arrived on the 16th. Back we went to Dunkirk, that time to load maize meal for Strood. By the 25th our 120 tons of cargo had been put into the holds and by the 2nd December it had all been taken out again at our destination.

We left for Bellamy's Wharf where 150 tons of cotton seed was loaded for Ipswich. It was quite a big cargo for us and needed a large stack on deck which required a lot of trimming and covering up before we set out. However we were ready by the 6th December and had arrived at Ipswich on the 9th. On the 15th we returned to London were we loaded maize for the Medway in the King George V Dock at North Woolwich.

Our cargo for Rochester was completed on the last day of December. It was our twenty-seventh of the year, more than a freight every fortnight, including two to the continent.

Bellamy's Wharf was in the upper reaches on the south bank of the Thames. Bellamy's Granaries on King & Queen Wharf were the only riverside bulk grain handling facility. It was a destination much frequented by sailing barges and motor coasters and was one of the last active traditional London River wharves.
(Courtesy of Museum in Docklands, PLA collection)

CHAPTER 11

Sailing Matches

"In my view all the other barges which failed to carry their boats should have been disqualified."

On the 6th January 1927, we made fast alongside the steamer Pancras lying at Woolwich Buoys and loaded a cargo of cotton seed. We arrived at Strood after a protracted voyage lasting four days and finished discharging on the 18th. Returning to the Albert Dock we loaded cotton seed again for Chatham and were subsequently back in the Medway on the 30th. The following day our discharge was completed and we were soon on our way to the King George V Dock where we loaded maize for Strood. It was not a large cargo, being only some 500 quarters, so loading did not take long and we were ready to sail on the 12th February. We made fast at Strood on the 16th and two days later we were on our way back to the Victoria Dock for our next cargo; short passages and regular work.

It turned out to be 1002 quarters of barley in bags which we loaded for Ipswich straight from the holds of the steamer Tuscaloosa City. Loading took place between the 22nd and 25th February, and it was during that time that I had an argument with the ship's clerk over the question of overtime. The customary working hours were from 8am to 5pm and there was no obligation to work after that time of day. It was however the practice for the ship's clerks to pay only one man per barge when overtime was worked. It was a practice that I had always opposed. After we had worked until 7pm on that occasion, I presented myself for the payment to the two crewmen, namely the mate and yours truly. After much argument with the ship's clerk I was unable to secure any satisfaction over the matter and I had to content myself with payment for only one person. As it happened the weather turned to rain which held up our loading. By 5pm on the following day we had still not finished and we were therefore needed again, but by then I was in a stronger position than before. After more argument I agreed to work the overtime, but only after being paid for the extra man for the previous evening, and for both of us for the hours ahead. There was never a doubt in my mind that on those occasions it was really a two handed job, especially with hatches to put on and cover up. It was the 28th February when we arrived at Ipswich with the cargo and we finished discharging by March 3rd.

We returned light to the Thames with orders to proceed up river to Dundee Wharf which was situated in the bight at the top of Limehouse Reach. We were sailing up on the starboard tack with our destination in sight when I decided we should brail up the mainsail. The mate took up his usual position at the winch to commence brailing the sail when suddenly he fell to the deck. I rushed forward as he struggled to his feet again. He complained of severe pains in his chest. Waving me aside he made his way forward to the fo'c'sle companionway and somehow managed to get himself down below. Bearing in mind that Limehouse Reach at half flood was one of the Thames' busiest spots, I had to get the sails off her and then try and get her alongside somewhere, all on my own. After slacking off the mainsheet and then running to the winch, I stowed up the mainsail, a little at a time, before dropping the head of the topsail. I rounded the barge on the tide, stowed the mizzen, downed foresail, clewed in the topsail, pulled in the vangs and, with a few spokes of the wheel to give her a sheer, I was able to catch a turn on some lighters and moor the barge. After heaving up the leeboards and making the barge safe, my thoughts then turned to the mate.

In the gloom of the fo'c'sle I found him in great pain, moaning as he lay on one of the lockers. He wasn't a chap to make a fuss. He agreed perhaps it would be wiser for me to go ashore and fetch a doctor. I managed to get a lift ashore and found one, but he refused to go out to the barge. We were close to Limehouse Pier and as it happened there was a little boat handy that I could borrow.

Limehouse Pier with the downstream part of Dundee Wharf behind. (Courtesy of Museum in Docklands, PLA collection)

Once back on board I struggled to get the mate, all sixteen stone of him, up the rungs of the ladder to the deck, then over the side into the boat, which was tossing about in the wash of many craft. It turned out he had broken a blood vessel just under the skin of his chest. He was black and blue all over and the doctor said he was to go straight to hospital. I found a taxi to take him and after a couple of days, surprisingly, he was discharged fit and well.

Our cargo from Dundee Wharf was potatoes. They were discharged at Strood by the 15th March. Returning light to Bellamy's Wharf, Rotherhithe, we loaded a big cargo, 193 tons 9 cwts of linseed. We set sail for Ipswich and arrived there on the 28th, finishing unloading by April 6th. After that it was back to London light, where we loaded a mixed cargo of grain in bags in the Millwall Dock for Strood. We unloaded there on the 20th and were ordered across to Dunkirk light ship. Arriving on the 22nd, we took aboard 130 tons of gluten meal which we carried back to Strood, the return voyage taking four days.

It was always a recognised thing to draw money on account before sailing, out of which the brokers dues at the various ports were paid and our food and necessities were purchased. Should any payment be made to me for cargo carried while away, then such sums were always sent intact, direct to the barge's owners. While these arrangements may sound plain sailing, there was one problem which caused quite a bit of hardship to the barge skippers. It was a problem that could, without very much ingenuity, have been rectified by a little thought from the owners. It was very seldom that a skipper could go to his barge owner's office and settle up without a lot of waiting around. Invariably there was some hold up, even when the skipper's precise time of arrival was

63

known in advance. Many times I telephoned the cashier the day before, telling him the amount of money that I would require and the time that I would call. Even so, on some occasions I would arrive in the office about noon as arranged and would be kept kicking my heels until about five-thirty.

All the skippers used to get furious at the time we waited, for this was often the only chance that we had for seeing our wives, homes and families. Many reasons were given for those frustrating delays; sometimes it would be the pressure of work, or else our waiting was while a visit to the bank was made. Then there was always that great favourite, no one available to sign the cheque. It seemed to me that if any spare time came the way of a bargemen, it was spent waiting idly at the owners office, but it was endured in those hard times rather than risking one's livelihood in an argument with the staff.

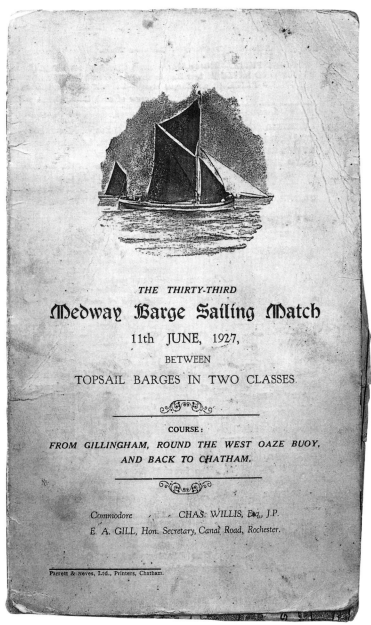

The 1927 Medway Barge Sailing Match programme. Note the order of words; not a sailing barge match, but a barge sailing match, a tradition which survives to the present day. (Albert Bagshaw collection)

THE THIRTY-THIRD

𝔐edway 𝔅arge 𝔖ailing 𝔐atch

11th JUNE, 1927,

BETWEEN

TOPSAIL BARGES IN TWO CLASSES.

COURSE:

FROM GILLINGHAM, ROUND THE WEST OAZE BUOY, AND BACK TO CHATHAM.

Commodore   -   CHAS. WILLIS, Esq., J.P.
E. A. GILL, Hon. Secretary, Canal Road, Rochester.

Parrett & Neves, Ltd., Printers, Chatham.

Our enthusiasm to earn a good living was the reason for the many good passages we made. If the work was there, then so were we. Bank holidays, anniversaries and bonfire nights would not hold us back, though we had to put up with the hardship of being away from home. But for some time the amount of work about had not been too good and so I decided that wherever we were bound, we would have to get by with just two of us sailing the barge.

On the 4th May we began loading manure and carbide at Barking for Newport and Southampton. We sailed on the 12th May and were ready for the trip back up Channel a week later. Although the run down Channel had made the barge £94, we had to bring her back to London light. We went to the Millwall Dock where a cotton seed cargo awaited us. We delivered 145 tons to Ipswich and on the 8th we were on our way back empty to Strood.

We were sent straight into the thirty-third Medway Barge Sailing Match, without preparation of any sort. I certainly didn't volunteer for it. Perhaps our performances of the previous two years had prompted the owners to enter her. These races never held very much interest for me for I considered that we were not fairly treated in competition with other craft. Generally speaking we could hold our own with any barge when it came to ordinary cargo carrying, but racing a barge was a very different kettle of fish. I had always held the view that it was the

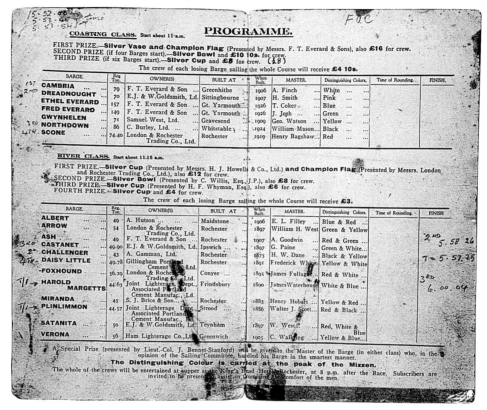

Eleven owners entered a total of nineteen barges in the 33rd Medway match. In the River Class *Plinlimmon* and *Harold Margetts* finished in a dead heat after six hours racing. The oldest and smallest competitor was the famous *Challenger* of 43 tons, launched in 1875. The programme details a supper for 'the whole of the crews' and that 'Subscribers are invited to be present to assist in prompting the comfort of the men.'
(Albert Bagshaw collection)

owners who should crew the barges during races, while the regular skippers and their crews should watch from the comfort of the Committee steamer.

The race was held on the 11th June 1927 and the course was from Gillingham down to the West Oaze buoy and back to the finish at the Sun Pier. The distance was nearly 35 miles. The race was open for two classes of topsail barges. The entry was made up of seven coasting and twelve river barges. *Scone* was the only London & Rochester Trading Company entry in the Coasting Class, though both *Arrow* and *Foxhound* from our firm sailed in the River Class. Although *Scone* was still only eight years old, there were three newer barges competing against us; Burley's *Northdown*, just three years old and two of Everard's new giants, *Ethel Everard* and *Fred Everard*, both around twice the tonnage of *Scone* and launched in 1926.

On that occasion *Scone* came in fourth in her class, with a time of about six hours for the whole course. We were strictly observing all the rules of the race, and accordingly as a working barge, we carried our boat. That was to be the subject of criticism which my owner made after the race. When he said "You should have won that race." I pointed out to him that amongst other disadvantages we had to bear, boat carrying was laid down according to the rules. In my view all the other barges which failed to carry their boats should have been disqualified. Few people knew that it was because of my protest over the matter that boats were carried in all the barge sailing matches from then on.

Later we sailed in another race but we were never a racing craft. You could not expect a hull built to stand all weathers at sea and to carry a big cargo to act like a racehorse. If the *Scone* had been fitted with racing leeboards and a sixty

The Coasting Class shortly after the start, with the chimneys of Gillingham in the background. L to R believed to be *Cambria, Fred Everard, Dreadnought, Northdown, Gwynhelen, Scone* and *Ethel Everard.* (S E Willis, Albert Bagshaw collection)

foot sprit, I am sure she would have been as fast as the famous *Northdown* and *Cambria*. It was their big sail area and special leeboards, together with three weeks or more preparation before the race, that made their reputations what they were.

Immediately after racing we went back to normal duties by loading timber up in the Surrey Commercial Dock on the 14th and 15th June for Chatham. It was reckoned that we had 32 standards aboard, but by comparison with the cargo which we had carried to Ramsgate earlier, it seemed more like 50 standards to me. Needless to say, we were only paid freight on the lesser figure.

After discharge had been completed on the 21st June, we went up the Medway to Holborough near Snodland to load cement for Penzance. Going up to the loading berth we had to pass under Rochester Bridge and that influenced me into shipping a third hand for the trip. The passage down to the Cornish port took us fifteen days or so, and through lack of foresight on the part of the owners we were forced to return empty. That unrewarding episode came close to making me say farewell to barge sailing for ever.

The first part of the voyage took us to Deal where we fell in with other barges including Everard's *Ethel Everard* and *Will Everard*, and their *Lady Mary* and *Lady Maud*. Also there was our firm's *Pudge* which was just five years old and the *Beryl* and *Scot*, plus quite a number of others.

From Deal we all went down to Dungeness together and then, taking advantage of the next tide, we passed on down Channel, leaving Hastings astern over on our starboard quarter, to Eastbourne where we all came to anchor for the next tide. The new tide was at two in the morning and when that time came round it was pitch dark and pouring with torrents of rain and there was no wind.

It was the lack of wind that led me to decide to wait until daylight. Returning to the modest comfort of my oil lit cabin I removed my outer garments and got into my bunk. As I lay there listening to the rain drumming away on the deck above I became aware of a drip, drip, drip. The heavy rain had found a weak spot somewhere in the flush fitting deck light above, but it took several years for me to trace the source of its entry. Back on deck around four in the morning I was surprised to find that all the other barges had gone, being helped to the west'ard by a nice steady southerly breeze that they had all noticed, but I hadn't.

With some forty miles to go to the Owers light vessel we lost no time in getting underway. By the time that we were passing inside the Owers we had already overhauled the tail-enders amongst the other barges. Shortly after we reached the Forts[1] the wind suddenly flew into the north-west and we had to make several boards to get into Stokes Bay where we anchored for the night. By that time we had parted company with all the other barges; most had made for ports like Southampton, Newport, Emsworth or Portsmouth to deliver their cargoes. By daylight the stragglers had still not made an appearance and in all probability they had suffered a dirty night and failed to get through the Forts.

We worked our way onward passing Christchurch Light buoy to Swanage Bay. There we spent the next night. When we mustered again the wind was blowing moderately from the west and we set out clear of the Peveril Ledge. We passed close by Anvil Point and then on down to St.Alban's Head. We made doubly sure that everything was covered up securely for the turbulent race, though it was not such a bad one as that at Portland. Then it was on to Portland harbour and finally into Penzance where we arrived on the 9th July.

The Harbour Master seemed none too pleased to see us. It turned out that a ketch barge, the *Martinet*, had just previously done some serious damage in the harbour. To cover the cost of repairs they had put sixpence on

[1] *The forts were part of a defence boom and fortification which was built to protect the eastern Solent in Napoleonic times.*

The *Martinet*, seen here anchored laden off Southend, was the last ketch barge in trade. She foundered in 1941 after stranding on the Whiting Bank whilst under skipper Bob Roberts. He eventually became the last barge master plying the traditional trades when he owned the *Cambria*, which he worked under sail alone until 1970.
(Tony Farnham collection)

the town's rates. We berthed in the outer harbour and by the 12th July had discharged our cargo.

During that time, had I been given a free hand, I could have fixed the barge for a return freight there and then. But the owners had lacked that foresight and I was compelled to wire them for orders. By the time their reply came back, giving me the authority to arrange the business, there was no cargo to be had. Neither was there anything on offer from any of the china clay ports or at Portland. Three weeks we lay at Penzance waiting for something to turn up but nothing did. Very frustrated we made the long journey of five days and four nights back to the Thames empty.

Off Herne Bay we let go the anchor for a good nights rest and felt a good deal better for it, before running up to the Victoria Dock in the morning. Here I met the owner's son, Maurice Gill, who had fixed our Penzance cargo without a thought of anything for the return. It was a turning point for me, because I'd made up my mind to come to a proper understanding with regard to the fixing of those long haul freights. In plain language I told him that if there were any more trips like the last one, with all the work and long hours it entailed and only a share of the freight on one half of the voyage as reward, then he could take the cargo himself in a wheelbarrow.

As a result we became almost a free-lance barge. I would be consulted before any long voyages were arranged so that I could endeavour to find a return cargo. That was far more satisfactory for me and the barge. I still had my worries though, for trade was still far from good. It started me thinking about the question of having an engine installed, converting the barge into an auxiliary.

At the Victoria Dock we loaded linseed for Ipswich, arriving on the 12th August. We lay there until the 26th when we were sent across to Dunkirk to load. Loading was complete by September 1st and we were at the Erith Oil Works discharging by the 5th. Again it was quite a smart voyage from Ipswich to Dunkirk, taking only two days. Orders then took us to the Albert Dock where we took aboard some 1600 bags of flour for Newport, Isle of Wight. Before sailing we went over to the Erith Oil Works again to top up the cargo with bags of pig meal. Loading was finally completed by September 12th and after a rather lengthy passage we reached Newport on October 8th.

Leaving Newport on October 10th we returned to the Thames light and made our way up as far as Gravesend before receiving orders to turn about in our tracks and go up the Medway under Rochester Bridge to Snodland where a cargo of cement awaited us. The cement was destined for Plymouth. I had fixed a return freight, even before we commenced loading.

When we had made that voyage down Channel on previous occasions with the tides unfavourable it was the custom to dodge to and fro in Weymouth Bay before rounding Portland Bill, or if the wind was fresh from the north-east then we would anchor in Portland Harbour. On that trip I tried a new idea. We stowed the foresail, clewed in the topsail and bent on an extra sheet to the jib, hauling it to windward. Then half the mainsail was taken up in the brails with the peak lines and the lowers and middle brails hauled nice and tight. In that way we were quite comfortably snugged down and were able to keep well clear of the Portland Race.

Once to the west of the Bill we let the barge come into the wind in order to hold the land and by the Portland light, which slowly changed from four flashes to three and then to two, I could easily see that the barge was soaking in quite steadily. We stood on and when daylight came we were in sight of Beer Head. That was the signal to set all our sails and we were soon running away down to Start Point and so on to Plymouth. We had taken a considerable amount of

Portland Bill and the Portland Race were notorious hazards to shipping. The SS Petroclus is ashore under the Bill surrounded by craft engaged to lighten her in the hope of refloating before bad weather puts paid to salvage. Two sailing barges play their part, the mulie between the rocks at the foot of the headland and the stricken ship has all her hatch covers removed and her main hold part filled with some of the steamer's cargo.
(Conway Photo Library)

water aboard during the passage and that was plain to see when we took the sheets off the forehatch. The water had found its way up under the fore end of the hatch cloth. Happily none of our 170 tons of cargo had been wetted. The freight on that occasion was 10/- per ton which brought in £85 for the barge, against which £13.6s.2d. gross expenses had to be met.

From Plymouth we went down to Par in Cornwall to load clay for Snodland Paper Mills. Arriving at Par during the fore noon of November 12th, we finished loading 169 tons six days later, the rate for which was set at 9/- per ton. Before we saw anything of the freight money however, expenses at Par amounted to £11.15s.10d. and by the time the cargo was delivered at Snodland late in November the gross expenses had risen to £14.9s.0d. The voyage up Channel to the Medway had taken us ten days, so in all we had only been just over four weeks making the round voyage. From Snodland we loaded 100 tons of cement for Limekiln Dock, Limehouse, where we were discharged by December 7th.

Directly after that was a passage from the Millwall Dock, which we took on December 10th to Ipswich, where we arrived on the 19th. After discharge we loaded drums of oil which we carried back to Hubbuck's Wharf, Ratcliffe. Leaving the drums at Ratcliffe it was down to the Albert Dock on the last day of 1927 to load wheat for Newport, Isle of Wight.

CHAPTER 12

**Auxiliary
Reluctance**

"... but not a minute too soon, for the moment we had made fast she sank under us."

The New Year had just begun and a strong easterly wind brought driving snow. With such awful weather I decided to accept a tow from the Pool down to the loading berth. The cargo was 750 quarters of wheat which carried a freight of 2/6d per quarter, grossing £93.15s.0d. for the barge. The easterly wind that was blowing as we loaded was a favourable one for the run down Channel, but not until we had rounded the North Foreland.

We came out of the Albert Dock lock and made our way down the Thames. That same wind prevented us from making any real progress and so I made the decision to anchor off the Ship and Lobster pub at Denton, just below Gravesend. We were forced to lay there for a fortnight, but it was nice to have the chance to get home for a few days and see Kathleen. I had a fairly easy mind as we had already made sure of a return freight of china clay from Wareham for Battersea. We finally left Gravesend with a strong north-west wind blowing, only to anchor again off Southend. The following day we ran over to the Ness Houses anchorage at the entrance to the East Swale, lying at the back of Warden Point. We spent a couple of days or so waiting there for better weather before running through the Downs and entering Dover harbour. The weather worsened and again we had to wait a few days more before venturing out and making our way down to Dungeness East Roads. Unfortunately the wind went round to the south and freshened considerably, driving us back the way we had come, with the added discomfort of a very nasty sea. With only the two of us to handle the barge, it was a very trying time.

Whilst we had retraced our course as far as Dover, I had made up my mind to go in through the eastern entrance. Unhappily I held out a bit too much to windward and lost the wind out of our head sails. We rammed the breakwater knocking out the bowsprit and damaging our stem. Nevertheless, we managed to get the barge under control again and got into the harbour, but with the wreckage of the bowsprit hanging over our stem I had to coax her very carefully. Because of the great difficulty that we were experiencing I decided to let her drift down to leeward and run into the Camber[1]. By that time the tug Lady Brassey had seen our plight and, with an eye to an easy salvage job, placed herself in a position which blocked the entrance.

It was only after I had sung out to him to get out of our way that we safely berthed the barge, and only then were we able to examine our damage in detail. The next day I hired a motor boat to tow us into the Granville Dock, and with a new bowsprit sent down to us from Rochester by the owners, repairs were got under way.

It was February 23rd when we finally arrived at Newport, again the first of several barges which had like us been bound down Channel from Dover. After discharging our wheat, which out-turned perfectly dry, we went over to Wareham for the return cargo. When loading was finished on February 29th,

[1] *A part of the dock complex at Dover, situated to the west within the outer harbour.*

The quay at Newport, Isle of Wight.
The Thames spritsail barge *Doric*
lays alongside.
(Tony Farnham collection)

leap year day, the wind had veered and blew strongly out of the north-east.
We had to turn away up to the Isle of Wight and with the tide we reached our
anchorage at Jack-in-the-Basket off Lymington. The next tide took us up to
Stokes Bay. We worked to wind'ard all day which meant the constant handling,
trimming and setting of sails; it was pretty exhausting work.

Aboard *Scone* the old mate, who was well past 65 years, slept for'ard and I
slept aft. We turned out in the early hours of the morning to light the fire and
make a cup of tea. While waiting for the kettle to boil we busied ourselves
getting the barge ready for continuing the voyage. We had anchored the night
before in six or seven fathoms of water, which meant that some thirty fathoms of
chain had to be hauled in on the windlass. Our departure was in the early hours
of the new day, around half past two, when most people were still asleep in bed.
It was, of course, dark and the mate lit the red and green side lights and took
down the anchor light, which was pulled up on a bight of the foresail halyard.
After lowering, it was stood on the fore hatch where its light would be helpful as
we pulled the foresail up above our heads so that the handles of the windlass
could be shipped.

As soon as the fifteen fathom shackle of the anchor cable came aboard,
which signalled that half the chain was in, I stood-by in readiness to let go the
topsail clewlines, having been aloft to release the gaskets which had stowed the
sail overnight. The sheet of the topsail was hove out and the head of the topsail
was lifted a few feet before letting go the mainsail peak brails. Then I
overhauled the main sheet block before letting go the lower and middle brails.
The main brails were then slackened off a little and the main tack pulled down

71

tight. The main sheet block was taken aft to be hooked on to the traveller on the main horse. The mate then slacked away steadily the main brail as the sheet was hauled tight home. Heaving on some thirty fathoms of three inch rope in the cold darkness before dawn, often with the mainsail thrashing back and forth across the deck, was no easy job.

Once the mainsail was set up tight, the rope sheet was coiled up clear and my attention turned to the rudder. Both the kicking chains[1] had been used to hold it fast after coming to anchor. They had to be removed in order that the helm could be worked.

[1] *The Scone was fitted with two kicking chains; one led to each quarter.*

The leeward leeboard was run down and the vang falls eased to allow the sprit some freedom and then the foresail was made ready. With a few turns on the windlass a short length of anchor chain was hove in, then the foresail was set to the bowline, and as the barge took a sheer the right way, the anchor was hove right up. During these manoeuvres to get under way, a sharp lookout had to be kept to avoid other craft.

We bowsed the anchor and then turned to setting the jib by first bringing down the bobstay. That also had the effect of tightening the long topmast stay and securing the topmast. After letting go the bowline on the foresail and raising the topsail the rest of its hoist, all our sails were set. All done, the welcome sound of the sea being sliced by the bows as she gathered way, sending the broken water bubbling and gurgling past the leeboards and down the sides of the hull, was something that never failed to excite me. At last a breather could be taken.

Under way with the tide behind us and not over much wind we sailed out through the Forts past Ryde and the Warner light vessel until we fetched up near the Owers when the tide finished and we had to anchor. As soon as the tide began to ease, we went through the whole performance of raising the anchor and getting under way all over again.

With the wind still north-easterly we worked our way up past Littlehampton, Shoreham, Brighton and Newhaven until finally coming to anchor again, to await another tide, off Cuckmere between the Seven Sisters and Seaford Head. After about five hours rest the tide eased and once again we got under way, laboriously working our way round Beachy Head, up inside the Royal Sovereign Shoal, past Eastbourne, on to Bexhill, Hastings and Fairlight, and finally coming to anchor in Dungeness West Roads. After the next change of tide we sailed on to St.Margaret's Bay and past Dover. The wind still came strongly off the land from the north-east and so we made sure that the fo'c'sle companionway was securely battened down and the fo'c'sle stove funnel was removed and well plugged. If the wind stayed where it was, then we were coming to the worst part of our voyage. We fetched up to the South Brake buoy marking the entrance to the channel between the Goodwin Sands and the Brake Sands. Passing the light vessel we got up above the Broadstairs Knoll buoy

The Brake light vessel laid to the south of Ramsgate marking the Brake Sand.
(Tony Farnham collection)

and by that time we were heading into some nasty seas. We had to drop the head of the topsail. The seas were coming aboard us from both port and starboard. The foresail was kept on the bowline to one side during one tack and the jib sheet to windward on the other. It was a question of sailing easy until the tide carried us far enough to wind'ard of the North Foreland.

While we were off Ramsgate the skipper of the barge *Royal George* sighted us as they came out of the harbour, and when we both met later after they had eventually arrived in London, he expressed surprise about the how the *Scone* had coped with the heavy seas that we had been taking aboard. Of course, once we had opened out Margate we went up the river like a race horse and finally arrived at the Battersea Works of Morgan Crucible on March 12th, fourteen days before any other barge that was coming up Channel.

We discharged our clay at Battersea whilst a lighter lay ahead of us was discharging plumbago or black lead. Both of these were dirty cargoes to carry, especially in wet weather, as it was on that occasion. The wind blew down from the lighter onto us, so the mixture of black lead, china clay and rain water made a filthy state of us. It was not often that types of cargoes clashed like that, but it meant that we had to have more than our usual wash down before we could even put the hatches on. After discharging at Battersea we were towed down to Woolwich having had to lower down again to pass under the bridges, just as we had when going up to unload. We went into the King George V Dock on March 15th where we loaded cotton seed for Ipswich, arriving there on the 23rd. After finishing discharging a week later we took on board 45 tons of oil in drums for two destinations, Rotherhithe and Ratcliffe.

By April 3rd we had unloaded and were sent to the Millwall Dock to load 144 tons of white cotton seed. Loading was completed by April 21st, the cargo very light in weight compared with its bulk, so that it demanded a good deal of trimming. In fact I think we trimmed it too tight, for I found myself wondering what was wrong with the barge when we came out of the Dock. She seemed very sluggish to handle. Nevertheless we arrived at Strood without incident on April 14th, but it was not until the 20th that we had unloaded.

Freights were very scarce at that time and we lay at Strood for virtually a month. We eventually went alongside Strood Oil Mills to load just 50 tons of oil in drums for Rotherhithe. Discharged by May 18th, we sailed from London light for Calais to pick up a cargo of about 150 tons of clinker cement for Cuxton Cement Works,

We arrived at Calais on May 21st after a good passage and were back and discharged in the Medway by the 31st. Then it was Millwall Dock again to load more cotton seed for Ipswich. That was out of her by June 11th and after a bare couple of days preparation on the ways at our shipyard, we had to take part in the annual barge match. We were still of no use at racing!

Immediately after the barge had completed the course for the race we sailed straight on up to Tilbury Dock and loaded 2,104 bags of cotton seed for Strood. After unloading on June 19th we went back up to the Albert Dock light where we loaded 2,400 more bags of cotton seed for Ipswich. Loading was finished by July 18th and discharge completed by the 25th.

From Ipswich we had orders for Rochester, but when we arrived we were sent back down river to Sittingbourne to load 170 tons of cement for Truro in Cornwall. After that tough last trip the two of us had down Channel I signed on a lad again as third hand. Before loading I enquired as to the availability of a return freight, for I had no intention of coming back light with all the work that was entailed and all the time such a trip could take.

From the cement works at Sittingbourne we came out of Milton Creek under tow, which incidentally cost us £2.15s.0d. covering both ways. It was almost high water as we swung to starboard, clearing the entrance of the creek. With a light westerly wind it did not take us long to gather way, once our tow had been slipped. With Elmley Ferry behind us we made our way down over the Grounds as they were called, out through the East Swale, down through the Gore and on to the Downs where we anchored for a tide off Deal. We got under way again at the break of day with a new tide, and by the time the light was fading we were at the Owers light vessel.

That meant that during the hours of daylight we had covered 95 miles from the Downs. Sailing through a starlit night with a nice breeze coming out of the east, we had run well with our square sail set. By daylight the next morning we were off Start Point and by the next evening we were in Falmouth. We had experienced an extremely good passage of only four days, reaching Truro on the 4th August and were discharged by the 9th.

In the heart of Truro the Ipswich built mulie, *Southern Belle*, lies alongside Lemon Quay whilst local trows lay opposite on Back Quay. Truro School dominates the hillside beyond.
(Royal Institution of Cornwall)

After discharging, we sailed out of Falmouth up the coast past Dodman Point to St.Austell Bay. We were bound into the awkward harbour of Pentewan where we loaded grit for the New Hythe Paper Mills. We arrived at Pentewan on the 11th August and secured a tow in and out for £2. We completed loading by the 14th and arrived back on the Medway at our destination above Rochester Bridge on August 22nd. By August 28th we had discharged our cargo of grit which was to be used to lay a 'marble' like finish to the new paper mill machine shop floor.

From New Hythe we went light ship up to London to load maize for Kings Lynn; an entirely new venture for me. By August 31st we had loaded 816 quarters of maize weighing 174 tons 17 cwts. Kings Lynn was certainly not one of the best places to go with a sailing barge. We took the usual course down to Harwich and then out to the Cork light vessel, keeping well clear of the Felixstowe Ledge, before shaping down to Orford Ness. Our course took us on past Southwold, Lowestoft and through Yarmouth Roads down past the Cockle light vessel and the Happisburgh light. After coming up to the Foulness buoy we swung round to the north-west leaving Cromer astern, shaping up to the Blakeney bell buoy.

I treated that part of the coast with great respect and took no chances, although in later years I grew quite familiar with it. Leaving Blakeney astern we came up to the Docking buoy where we turned onto a course up to the Lynn Well light vessel and so on to the amusingly named Roaring Middle light. From here we entered Lynn Roads, where for my first visit I was happy to find it a compulsory pilotage area.

We were only able to reach the bottom of Lynn Cut before coming to anchor. We laid there all night and when the Pilot came aboard the next day he wanted us to pay for a tow up the cut. I was not going to have that when we had a fair wind. The outcome of that difference of opinion was that our voyage was concluded by us sailing up. After discharging it proved very difficult to secure a return freight. At last I was successful in fixing the barge to lift coal from Goole for Shotley, near Harwich. We went on down to the Humber and were loaded by September 18th. We were ordered for Shotley Hard and when we arrived there on September 21st the tides forced us to wait for water. We finished discharging on September 29th and were then ordered to Sittingbourne to load cement for Poole.

After sailing through the Kings Ferry Bridge we were fortunate in securing a tow up to Sittingbourne where we commenced loading late on October 4th and finished the following day. Out trip turned out to be a protracted voyage, for after leaving Sittingbourne and proceeding down over the Grounds we lay wind bound for a week in the East Swale off Harty Ferry. There were quite a number of other craft in a similar predicament. We were in company with the ketch barge *Clymping*, one of four boomies which later sailed to the West Indies and worked in the sugar trade for many years. The others were the *Goldfinch*, *Kindly Light* and the *Leading Light*.

Finally, we were able to leave the Swale only to become wind bound again for several days at Dover. In spite of the old proverb 'When the wind backs against the sun, you can be certain, back it will run', when the wind did finally back against the sun, to the south-east, I put to sea. It was not entirely bad seamanship on my part, for I had always been under the impression that if the wind backed before noon it would last at least twelve hours.

With a strong south-east wind filling our sails we left all the other craft at Dover and, as was my customary good fortune in such matters, I got a good offing and carried that south-east wind almost to the Owers. Here the wind began to veer, but our offing had been too good for it to cause concern. By the time the wind was blowing from the south-west we had rounded the Owers light vessel and were heading in for the Wight. Spending the night under Ryde, we went down to the Jack-in-the-Basket anchorage the following day. The strong westerly wind compelled us to lay at our anchor for a couple of days or so. Eventually we arrived at Poole on October 23rd, but as the weather was showery our discharge did not finish until the 27th.

The return trip to London had to be made empty for there was just nothing to be had. In spite of the continuing bad weather I decided to put to sea and make for the Isle of Wight, which was not a great sailing distance away. A light barge made a lot of leeway, so we left Poole holding to windward close to Old Harry and the Swanage buoy until sighting Anvil Point, which was the signal for us to bear away for the Needles. A nasty sea was running the whole time, and it was already dark by the time we fetched up to there. As we ran in between the Needles light and the Shingle Bank buoy, the barge suddenly dropped her stern so deep in a trough that I thought our bowsprit, which was still topped up, would fall inboard! The next moment a heavy sea came in over the stern and I found myself desperately hanging on to save myself from being washed off the barge. The wave passed and soon we were inside Hurst Point coming to anchor off Yarmouth, where we stayed for the remainder of the night.

Next day we moved our anchorage to Ryde, for the weather seemed too bad to go on any further with our voyage. The following day opened very much the same, a strong south-westerly wind and a very heavy sea. Nevertheless I decided to carry on. Holding as much to windward as possible and with the barge heeling right over on her side for most of the time, we drove up Channel under reduced canvas.

From the Nab Tower we shaped away to weather the Owers light vessel from where we had to keep a good offing in order to make certain the weathering of Beachy Head. It was not long before we arrived in the Downs and Gravesend was reached on November 1st, so I was able to get home, and the third hand left us.

Tilbury Docks were *Scone's* next loading place and by November 6th we had left for Strood which we reached on the following day. Returning up the Thames again we loaded 120 tons of rag-stone from a steamer lying at Cory's Wharf, which we carried to Sheerness Pier. Loading on November 14th, we berthed at Sheerness on the following day in a heavy south-west gale.

We were accompanied by the barges *Louise* and *Milton* which both loaded with us. They anchored off Sheerness Pier to await their turn to berth and while so doing the *Louise* was swamped as she lay at anchor and sank, while the *Milton* dragged her anchor and drove into the pier. We went to her aid, taking a heaving line out along the pier. After quite a struggle we were able to get the skipper and mate to safety.

Later as the tide ebbed I managed to board her, only to find she was about one third full of water. The barge had damaged herself badly and had been leaking. By bending a line onto the plug I managed to pull it free and empty her before the tide came again. With the next flood the weather began to moderate. Working her pumps we wondered if the leak could be held. We managed to get the *Milton* brought properly alongside the Pier, but not a minute too soon, for the moment we had made fast she sank under us.

Eventually she was discharged, and with the help of my mate, we took her round to Rochester where we had much difficulty in making her owner appreciate the amount of effort we had taken to keep his barge from becoming a total wreck.

Shortly after that little episode I was asked to go to Groningen to collect a motor ship which I brought back to Rochester, along with a very tall Dutch engineer. I took him home to meet the wife, where he had to duck to walk through our doorways. For some time I had been agitating for a powered craft, but that motor ship did not suit me in the least, for I could see from my short passage in her that she was liable to give a lot of trouble and would most certainly fail to pay her way. My own idea of motor power was to convert the *Scone* into an auxiliary, so I refused to take the newcomer and stayed in sail.

We took the *Scone* up to the Surrey Dock where we loaded timber for Faversham. By December 20th we had only about 25 standards and it was not until Christmas Eve that we were at our destination ready to discharge. It was early January 1929 before we were unloaded. With the New Year only eleven days old we entered the Victoria Dock, where we loaded linseed for Ipswich. By January 23rd the cargo had been discharged and we returned light back to London. We passed up through Tower Bridge and loaded in the Upper Pool for discharge at Thurley's in Wandsworth Creek. That was our first work above bridges since March of the previous year, with our gear to be lowered and raised to discharge and again to come down river.

It was early February when we were ordered down to Dundee Wharf to load potatoes for Margate. Loading took two days and the weather was bitterly cold. When we arrived off Margate Roads we came upon the barge *T.T.H.* anchored under the lee of Margate Sands and unable to raise her anchor. Her windlass had frozen solid with the spray which had cascaded over it. She was forced to remain there until it had thawed out, her crews' efforts with hot water from their stove making no impression.

It took us a couple of days to discharge and then it was back up the Thames light, through Tower Bridge to Mark Brown's Wharf in the Pool of London where we loaded for Ipswich. After arriving at Ipswich on February 22nd, and finishing our discharge on the 28th, we returned light to the Millwall Dock to load wheat for Shoreham.

Whilst waiting to load I had an offer to take command of a 1,000 ton steamer, with a wage of £8 per week, and a navigating mate. My heart was still in the sailing barge, so I turned that offer down; the second powered vessel which had been offered to me and the second I had refused.

With our many comings and goings we had to keep a careful eye on the stores which were so essential to our job. There was always the question of buying food for example, the provision of fresh water, paraffin oil for the lamps, coal for the galley stove and skipper's cabin. Firewood for these was not such a

problem as there was always enough driftwood to be picked up from the water to satisfy our requirements, but nevertheless it had to be dried out before being sawn or chopped up ready for use. Then of course, there was the all important task of keeping an adequate supply of emergency rations aboard for when unexpected spells of gales, calms or whatever kept us from obtaining fresh supplies.

March 4th saw the last of the Millwall wheat cargo go under the hatches, and Shoreham was reached safely and with a dry cargo on the 12th. After discharge we commenced loading scrap iron in the Sussex port for Antwerp. Coming out of Shoreham on the 18th, we had a good sail to the Downs, where we waited for a while for the tide and the wind to turn in our favour.

We then headed for the South Goodwin light vessel and had only just cleared her when fog arrived, together with a fresh south-west wind. We groped our way to the Ruytingen light vessel, her fog horn the only clue we had of her whereabouts. We sailed on to the Wandelaar light ship which I approached with some anxiety, for a ship had been sunk in the vicinity and we had heard that the barge *Solent* had hit the wreck and foundered. As we passed the danger area with the fog still thick and the wind still fresh, around us the air was echoing with the sirens of numerous steamers struggling in the murk.

After shaping in towards Zeebrugge we passed up along the Belgium coast towards the entrance to the River Scheldt. Rounding Nieuwsluis Point we dropped anchor before resuming our voyage next day up to Flushing where we took on the Pilot for the trip to Antwerp. We finally arrived at our discharging berth on March 25th with our 170 tons of scrap, for which the freight was 5/6d per ton. Our expenses during the voyage totalled £10.4s.0d. which resulted in a net freight of £36 which was as usual split equally between owner and crew.

Whilst we were there our ship's cat, Ginger, disappeared. He was always ashore as soon as our mooring ropes were, but I only had to call his name for

Deptford Creek had probably changed little since this picture was taken sometime before WWI. The first vessel owned by F.T.Everard was the sailing barge *Industry*. Although around a dozen sailing barges bore the name, this might be her in the left of the picture.
(Tony Farnham collection)

him to bound back aboard; but not that time. We called and searched but without success. Then a passer-by suggested that they ate cats in Belgium, and we would not see him again and I came to terms with his sad demise.

There were several craft waiting at Antwerp for a return cargo to England. Bearing in mind the experience I had in Penzance, I took the first cargo that was offered. That was a consignment of bricks from Niel, which was well above Antwerp, needed for Deptford Creek

Ginger, *Scone's* cat, was banished ashore after going absent from the ship on a trip to Belgium. (Albert Bagshaw collection)

at 5/- per ton. Just as we were about to let go, Ginger came running along the quay and leapt aboard. I decided that he should swallow the anchor and when I got back home I made him a nice box alongside the kitchen stove. In his final days on earth that old cat just lay there awaiting my return from a long trip away. Moments after I got to see him and say hello, he passed gently away.

Leaving Antwerp on April 2nd we sighted the wreck we had missed in the fog on the passage out. We arrived in Deptford Creek on April 11th. Having delivered exactly 172 tons, the gross freight came to £43, which after our various expenses had been deducted and the owner had taken his share meant that we sailed a barge load of bricks from Antwerp to London for just £10.

After that long passage we made the short run from the Albert Dock to Strood with cotton seed, returning with 110 tons of monkey nuts to Rotherhithe. After these two river freights were completed on April 27th some of the craft we had left behind at Antwerp were still there, waiting for a better freight. On May 1st we had again loaded linseed in the Millwall Dock for Ipswich, where we arrived on the 6th. That was followed by another identical cargo for Ipswich which we discharged by the 24th May.

By way of a change we were next ordered to Goole to load coal for Gravesend. Before leaving for the Humber we came down to Shotley Spit where we became windbound by strong easterlies. While we lay there, I kept my ears open and managed to find and fix a freight of shingle from Orford Haven up to New Holland in the Humber. That was new ground for me, and when we entered the river to load we found that we had to take our cargo off from the shore in wheelbarrows. Surprisingly it was more quickly achieved than one would imagine and we were all battened down by the 5th June. New Holland was reached on the 11th where our discharge was completed the following day. The freight was 4/9d per ton for the 182 tons we had loaded, and our expenses came to £8.

Although not a drinking man, while we were at Orford Haven, I accompanied the mate ashore to the local pub. The crew of another barge, with whom I was not acquainted, came into the pub and started to tell an incredible story of when they lay windbound at the Shore Ends anchorage in the Crouch estuary. They told how after some days they were getting short of food, so it was decided they would go ashore at dusk to get a sheep, perhaps some swedes and potatoes. So accordingly they first donned their knee boots, brought up their little boat alongside and away they went pulling the boat into shoal water, where the skipper and mate got out to lighten it. Dragging their boat through the shallow water for the last few yards they trod on some flat fish, which they grabbed and put in their boat. They went on to tell how they got their boat ashore and climbed up the sea wall stumbling over and landing on some partridges, so they wrung their necks and took them. Going down the other side they had to cross a ditch where their boots filled and got stuck, so they decided to leave them until they returned. Climbing up the other side they fell again, this time on a pheasant or two, so their necks were wrung as well. Then they made their way towards the swedes and potato clamps. While fumbling around for the swedes and potatoes they disturbed a hare which ran straight towards them and was killed and put in their bag. They collected the swedes

Charter Party. ~~Plymouth~~, Gravesend dated 26th May 1929

It is this day mutually agreed between Capt Bagshaw of the good Ship or Vessel called the *Scone* of Rochester of the Register of Ipswich Tons, all carrying 170 Tons or thereabouts, now lying in the Port of Gravesend and W. D Crook (Seaborn Coal Co) Merchants, that the said Ship being tight, staunch, and strong, and every way fitted for the Voyage, shall with all convenient speed, sail and proceed to Goole

or so near thereunto as she may safely get, and there load from the said Merchant or Agent, in the customary manner, a full and complete cargo of House Coal

not exceeding what she can reasonably stow and carry over and above her Tackle, Apparel, Provisions, and Furniture, and being so loaded, shall therewith proceed to Marrits Wharf

or so near thereunto as she can safely get, and deliver the same to order of the said Merchant or his Assigns in the customary manner on being paid freight at and after the rate of Six Shillings per ton and 30/ gratuity to the Captain

(The Acts of God, the King's Enemies, Restraints of Princes and Rulers, Fire, and all and every other Dangers of the Seas, Rivers, and Navigation of whatever nature soever, during the said Voyage, always excepted). The Freight to be paid on unloading and right delivery of the Cargo by the Consignees in Cash, Ship to be loaded and discharged with all possible dispatch

to be allowed the said Merchant for loading the said Cargo (to commence when Vessel is ready to receive Cargo), and for discharging. If required, the Vessel is to lay 10 days on Demurrage at £ per day for every day of detention, over and above the said laying days (except occasioned by Riot, Strikes, Frost, or Floods, or other accidents which may prevent the loading or delivery of the Vessel). Penalty for Non-performance of this Charter Party, estimated amount of Freight. Commission at the rate of 5 per cent. is due on signing this Charter to W. D Crook on Freight, Dead Freight, and Demurrage, and the Vessel on her return to PLYMOUTH, to be addressed to him, or to his Agents at Ports of loading and discharge. The Vessel to have an absolute lien on Cargo for all Freight, Dead Freight, and Demurrage.

Witness (Signed) W D Crook (Owner or Master).

Witness (Signed) by authority of parties (Merchant or Agent).

(vertical note): This Charter to be continued for next round voyage as from this date. W. D Crook, 21 st June 1929. S Bagshaw

Printer: Rt. White Stevens, Plymouth. 09685

This Charter Party form was printed for charters made in Plymouth but has been altered by hand in ink for use in Gravesend. It is for *Scone* to carry 170 tons of 'House Coal' from Goole to Gravesend at 6/- per ton and specifies a 30/- gratuity to the Captain, and that the 'Ship to be loaded and discharged with all possible dispatch.'
(Tony Farnham collection)

and potatoes and then caught a sheep which they carried back to the ditch, by which time they found their boots were full of eels! The publican interrupted "You'll do, here's your beer." When we enquired we learned that it was the practice of the pub to give a gallon of beer to the person who could tell the biggest lie on a Saturday night!

On June 13th we loaded 162 tons of coal at Goole for Gravesend, the freight being 6/- per ton with expenses of £11, which included the hire of a tug to tow us from Hull Roads up to Goole and back down again. Later, when that became a regular run, it was the necessity of incurring the expense of the towage that strengthened my argument for fitting an engine into the barge.

We arrived at Gravesend on June 19th and finished three days later. From June 25th to the 27th we visited four different wharves in the London area to take on a cargo of empty oil drums for Ipswich. On July 4th we went down light to Goole where we again loaded coal for Gravesend; a cargo of 171 tons. We were back at Gravesend on July 19th. After discharging on the 24th we went up to the Surrey Dock to load timber for Faversham. We took on 37 standards at 18/- per standard. That was as much as we could carry, even with a sizeable stack above the hatchways, and by the 7th August we had completed the trip.

This large mule rigged barge has taken the ground in the River Medina below Newport, Isle of Wight, and awaits the tide. (Tony Farnham collection)

The next freight was a new cargo for us. We were to carry asbestos slates for Poole. A total of 128 tons was loaded, and we still had room for more, so *Scone* was topped up with 48 tons of oil cake for Newport, Isle of Wight. Loading took place between the 12th and 14th August, the asbestos slates being taken aboard at Erith and the oil cake at Strood on the Medway. We got to Newport on August 27th and by the evening of the following day we had discharged the oil cake and made our way over to Poole, arriving on the 30th August.

By September 4th we were again ready to receive a cargo. For that we had to go further down Channel to Par, where we loaded china clay for Snodland on the Medway above Rochester Bridge. About that time there were a couple of ketches trading regularly to the Thames, and until that voyage I had always held the impression that they were smart little craft. On that occasion, while off Plymouth, bound up Channel, we fell in with one of them. That night the wind blew from the south and as was my usual practice I held well off the land. When daylight came next morning I looked about expecting to see her well ahead, instead of which she was well to leeward and astern of us. We managed to get inside the Wight, whereas later when we fell in with her again, the skipper said he had only just managed to make Portland. On that voyage our cargo had amounted to 170 tons which earned a freight of 8/4d per ton. Our expenses came to £15.10s.0d. leaving a net freight of about £57 which meant £28.10s.0d for the owner and the same for the crew.

Our next voyage was from Tilbury Docks to Ipswich with cotton seed, and then up to Goole to load coal for Gravesend. By the time we had finished discharging the coal it was October 23rd and from Gravesend we went up to the Surrey Dock to load linseed for Northam near Southampton. Linseed was a cargo that had to be stowed properly so that it could not move, for it ran like water and movement of a cargo when at sea could be very dangerous. Dockers were usually completely oblivious to the perils of our everyday lives, taking the barge down Channel in winter time, indifferent to the need to work the cargo up to the deck in the wings of the holds. That we had to do for ourselves and after much hard work, our efforts gave us peace of mind. We finally loaded the 150 tons with a good deal of bad feeling and language, and set sail on October 31st.

By the time we reached Dover a south-westerly gale forced us to anchor in Dover harbour. Before darkness fell the wind was blowing with gusts of up to 90 miles per hour. We were lying to two anchors, one with 75 fathoms of chain out and the other with 30 fathoms out. Because of the poor holding ground which was mainly silt, every craft, with the sole fortunate exception of ourselves, dragged their anchors in the night. The Harwich boomie *Mystery* blew up onto the beach and became a total loss.

With a living to earn not only for myself, but for a wife and our first child I did not want to hang about in harbour. At the first slant of a favourable wind we hove up both anchors and put to sea again with the benefit of a south-easterly breeze. We hadn't quite made the Owers light vessel when the wind swung round to the south-west again. With our lee deck under water we reduced sail in the pouring rain. Making our way inside the Horse Sand Forts, a coasting steamer hailed us to enquire about the weather. I told him that it was bad and that we intended to anchor under Ryde for the night. We had sustained no damage and we had no pumping to do, but we were wet through and exhausted.

It was on November 19th that we finally got up to Northam. With a gale still blowing and heavy rain falling I went up to the mill office to find out about the discharging prospects. The mill manager was unknown to me and I learned that he had later written a letter to our owners expressing the view that I had some audacity in expecting to be discharged in such weather. However, we were empty by November 23rd.

It was becoming increasingly difficult to earn the good living that we had been accustomed to. That linseed freight had brought in just £60, against which we spent £10 expenses.

It was during that stay at Northam I first had an opportunity of seeing some Kelvin engines, and they impressed me so much that I wrote to the makers for the fullest information concerning them. All that information I later presented to my owners, although I knew full well that the prospects of getting one installed in the *Scone* were rather remote. I had already by then turned down one full powered motorship and an auxiliary motor barge, both offered to me by our firm.

The little known port of Northam, up the River Itchen at the top of Southampton Water. The *British Oak*, seen alongside, was built at Maidstone in 1903 and gained 2nd and 3rd places in the Thames Match in her early years.
(Tony Farnham collection)

The latter was by that time having the engine taken out of her as she would not pay. She was up for sale and did not help my proposal at all. Later I learned that she had been built with the engine in her, and that the propeller had to deliver within the area of the thick rudder post. That reduced the barge's efficiency under engine and also meant that she would not sail properly. That gave some strength to my argument that engine or not, the vessel must be efficient under sail. Another little motorship which I had looked over before dismissing her as unprofitable was a 170 tonner which had been bought by a Greenhithe firm, but for my purposes I considered her too small.

Trade was particularly quiet at Southampton and the best offer that came our way was a little freight from the Southampton Docks up to Northam with cotton seed. After we finished discharging on December 3rd we went over to the Medina Cement Works between Newport and Cowes to load cement for Weymouth.

We arrived in the River Medina on December 4th and that night it blew so hard that we had to put our eight inch tow-rope out to hold us alongside the cement works wharf. Although we managed to load our 160 tons in little over a day, a heavy south to south-westerly gale forced us to stay put until the weather eased. We could not lay off the wharf as there were too many local craft sheltering there, so there was nothing else for it but to move out of the river and make our way to Osborne Bay, below Queen Victoria's favourite residence, Osborne House. There we dropped our anchors and

Two barges, one a mulie, lay
unloading in Weymouth harbour.
(Tony Farnham collection)

were forced to lay idle for several days. As a result we did not manage the short passage to Weymouth until December 16th.

When our discharge had been completed we went round to Portland to load stone for Cubitt Town, London. After we had loaded, the weather was still bad, so I arranged with the Harbour Master at Weymouth that we should return and lay there over the Christmas. It was not until the last day of the year that conditions permitted us to put to sea, but we were hardly up to St.Alban's Head when the wind suddenly flew into the south-west again. Bearing in mind my previous experience off the Needles Channel, I decided to keep well off and steer a straight up Channel course in spite of the heavy seas. Believe me those seas were heavy, and they made the barge pitch awfully as they came up from behind us. The stern would drop into the troughs and the crest of a wave would break right over the poop so that I was often up to my waist in water while standing at the wheel. When we were a little above Dungeness the wind veered to the north-west but we carried on up under Broadstairs where we anchored. Once again we were tired and wet through and through.

Next day the wind looked to be going further round to the north so we got under way with the intention of making Pegwell Bay. As we came up between the Brake Sand and Ramsgate the foresail blew to pieces when we were somewhere in the vicinity of the Old Cudd Channel. There was some five fathoms of water under us, so we let go the anchor and stowed up the sails. It was not long before boatmen from Ramsgate put out and came aboard with all manner of suggestions; if we stayed where we were we would break up or go ashore and so on. They might have been local men and quite likely knew the area well, but their pessimism didn't rub off on me. I told them to wait and see, for I knew what I was doing and made up my mind to report being pestered by these men to my owner, as indeed I did.

That night the weather began to improve and we got under way again, working our way up to the Spit buoy where we spent another night at anchor. Next day I sighted the little motorship which I had brought over from the builders in Holland, and after exchanging signals he towed us up to Sheerness. Here a new foresail was sent down to us and we finally made our destination at Cubitt Town on January 10th 1930.

The year of 1929 had seen us carry twenty-three cargoes. From my own records I could compare our efforts on the *Scone* with those of the new motorship. Whereas the latter had made forty two passages with cargo, against our twenty three, in spite of the poor trading conditions we had done better financially.

Our cargo had been discharged at Cubitt Town by January 14th, and the next was 1,600 bags of ground nut meal which we loaded at Erith Oil Mills for Strood. We discharged by January 22nd and returned to Millwall with 161 tons of ground nuts. Then we went across to Dunkirk light to collect a cargo for Strood which was discharged by the afternoon of February 15th. Trade was going from bad to worse, and we laid at Strood for ten days waiting for orders; no work, no money coming in. That period did however coincide with the birth of our second son Albert on February 17th. Unlike the arrival of his brother Cyril, I was able to be home for the happy event. Eventually we went back up to London empty to load maize for Kings Lynn. We discharged on March 10th and left light on down to Goole where 163 tons of coal was loaded for Gravesend. When the coal had been delivered we carried linseed to Ipswich, and then loaded shingle again at Orford Haven for Grimsby. We arrived there on April 15th and had discharged a couple of days later.

Many barges lay idle hoping for work. Goldsmith's ironpot *Speranza* and the *J.B.W.*, which was mined with the loss of her crew in WWII, await with others for a freight. (Conway Photo Library)

That was followed by another trip up to Goole for a cargo of coal for Gravesend. It was a voyage that had a rather unfortunate ending, for as we came slowly up river, weaving our way in between various craft towards the discharging berth at Gravesend, the tug Crusader, which had been heaving its anchor up, came ahead and collided with us damaging our starboard bow. The accident ultimately resulted in a law suit for which I duly appeared in court. The immediate result of the collision was that we had to go to our Frindsbury barge yard at Strood for repairs, arriving on May 2nd. The opportunity was taken to unrig the barge for a general overhaul whilst the repairs to the bow were being carried out.

Having had a bad trading year that far, I again brought up the question of an engine, without which I could not see a proper living in the future. Again I met with a refusal. The Director's view was that the lamentable trading conditions did not make the installation expense worthwhile. I told him that the meagre and irregular freights were hitting me at least as hard, if not harder than the firm.

The owner's refusal was followed by a request. Would I sail another barge in the coming barge race? Feeling as I did about earning my living first, I refused. Then a rather heated argument developed between us which resulted in my decision to leave the barge and the company before the week was out. Just before my departure I was offered an engine for the *Scone*. As I already had another job to go to I did not change my mind about leaving, although I did change my mind and promise to sail in the coming barge race. At the owner's expense we were kitted out with matching jerseys for our racing crew, spending £3.9s.0d. on our knitwear, but unfortunately the result was no different than before.

CHAPTER 13

Paying
Proposition

"We had carried 35 cargoes, against the 24 which we would have typically carried under sail alone."

After the barge match the owners offered me a freight to Poole as a bait! But I steadfastly refused to return to the company unless the promised engine was installed in *Scone*. Accordingly, I went back to my new job ashore, employed in charge of a large wharf in West Street, Gravesend, owned by the Gravesend Seaborne Coal and Wharfage Co. Ltd. However, my stay ashore was short, for just two weeks later I received a letter from my previous employer.

The contents of that letter and the events that followed brought about a new era for the *Scone*, changed from total reliance on wind and tide to a more controlled method of journeying around the coast. Yet, as I looked back over the remarkable voyages I had made under sail, often just with the two pairs of hands aboard to control that huge area of heavy canvas, I could not help but feel a sense of achievement and nostalgia. As a result of incredibly long hours and very hard work in all weathers, we had run almost as regularly as any steamship.

Despite his brief stay, Captain Bagshaw had obviously impressed his new employer.
(Albert Bagshaw collection)

Gravesend & District Seaborne Coal Co.,

GRAVESEND SEABORNE COAL & WHARFAGE CO., LTD.,

MARRIOTT'S WHARF, WEST STREET.

Registered Office—
3, OVERCLIFFE.

Telephone 524.

Branch Order Office—
31, PARROCK STREET,
and all
AUTHORISED AGENTS.

THIS IS TO CERTIFY that Captain H.W. Bagshaw was in the employ of the Gravesend Seaborne Coal and Wharfage Co., Ltd., from 2nd June, 1930 as Craneman and Wharf Foreman. He is leaving the employ of the Company only because he has been asked to return to his previous occupation as Master of the barge "Scone" having been offered more advantageous terms by his previous employers The London & Rochester Trading Co., Ltd.,

I very much regret that my Company is losing his services as Captain Bagshaw has proved himself a thoroughly trustworthy and conscientious Chargeman. He has worked hard in the interests of the Company and given excellent services in every respect. He leaves with the best wishes of the Company for his future welfare.

Gravesend Seaborne Coal & Wharfage Co.,Ltd.,

Wm. D Crooks.
Director.

12th July, 1930

The owners had decided to accept my advice and fit an engine. It would be a 30 horse power Kelvin and the work of installing it was soon to be completed. With that offer I gave notice that I was going to return to the barge. Although my new employer didn't want me to leave he was quite understanding about the circumstances and wished me well.

Except for the removal of the squaresail, the barge's gear was exactly as before. Below, some of my cabin space had been taken away to form part of the new engine room.

It was July 10th 1930 when I returned to the *Scone* and on the 18th we commenced loading timber from up in the Surrey Commercial Dock for Ramsgate. We were not finished until July 31st, but Ramsgate was reached on August 2nd and we completed our discharging by the 6th. As the engine was not working entirely satisfactorily we took her back to the yard at Strood for adjustments to be made.

By the 9th August we had begun to load 170 tons of bagged cement for Southampton from the Rainham Cement Works. Loading was finished a couple of days later, after which we put to sea and arrived at our destination on August 20th. Before the discharging had been completed we received orders that we were to proceed to the Medina Cement Works, over on the Isle of Wight. There we were to load for Gweek on the Helford River in Cornwall. It was on the 27th August that we sailed from the Wight. The passage to Gweek was made in two days, arriving there on the 29th.

The Helford River and its surroundings were very peaceful. Or they were until a boatman came out to us complaining about how we were damaging the Prince of Wales' oysters. I drew his attention to the fact that we only drew six feet of water, and that there was at least twelve feet of water under us. Still muttering, he headed back to the shore. Later the pilot came aboard and took us up to our unloading berth, passing Frenchman's Creek on the way, made famous in Daphne du Maurier's book. Discharging was completed by August 29th and we sailed to Charlestown to load china clay for the New Hythe Paper Mill at Aylesford above Rochester Bridge. We loaded between the 4th and 5th of September, and had an uneventful return trip to the Medway. Our discharging was completed by the 10th, which meant that we had been away down Channel for exactly one month.

The cement cargo of 170 tons had been carried from Rainham to Southampton at a freight rate of 6/9d per ton and the expenses came to £9. The 130 tons of cement to Gweek carried a rate of 8/3d per ton less expenses of £14. These expenses were considerably more than usual as we had to pay for the discharging. Finally the cargo of china clay was carried at 7/6d per ton with the expenses working out at £8.12s.0d.

The installation of the engine had resulted in a re-arrangement of the working and pay conditions between the owners and myself. I had agreed to stand the cost of the fuel oil out of the crew's share of the freight, while the owners were to provide all lubricating oil and maintain the engine out of their share. Freights at that time were so low that our first four weeks with the engine resulted in the crew only earning about £45 from which the fuel oil costs had to be deducted. I began to wonder if I had done right in returning and wondered again about giving up the sea. I had a wife and two sons to keep at Northfleet and myself aboard the *Scone*. That was like keeping two houses going at the same time. Even when I found the time to travel home for a few hours, there was still the worrying cost of the fare.

The next freight was 750 quarters of barley from Ipswich to Hull. It was a long trip down for only a few shillings. The cargo came to about 150 tons at 5/- per ton, giving gross earnings of just £37.10s.0d. However we took what we could get to keep the barge working as I had to accept the responsibility for proving the installation of the engine was a paying proposition.

While we were northbound with that cargo we had to put into Great Yarmouth because the barge had begun to leak under her runs aft[1]. The reason was simple; the barge had been built with caulked seams and the vibration of the engine had shaken the caulking out. That was soon attended to by local shipwrights and there was no damage done to our cargo.

Our Kelvin was a two cylinder unit, running on paraffin but starting on petrol. By having it installed on the port side, so the shaft went through the port quarter of the stern, the necessity to bore a hole through the thick stern-post was avoided.

[1]*The parts of the underside of the barge where the bottom planking sweeps up to the transom.*

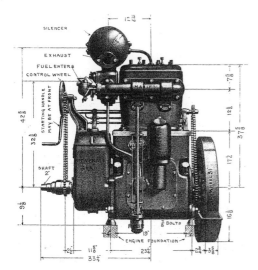

The two cylinder 30 horse power Kelvin C2 engine installed aboard *Scone* in 1930.
(The Ballast Trust collection)

The engine had 6 inch pistons with a stroke of 9 inches which developed about 30 horse power. It gave us a speed through the water of about four knots. It was hardly powerful enough for the job and in any tideway I had to use a good deal of discretion to keep out of trouble by using the sails.

On September 17th we arrived at Hull with our cargo of barley which we discharged in the Old Harbour as it was called. Running right up the river as far as Stoneferry meant negotiating no less than seven swing bridges as well as other craft. As the stream's width was too narrow to swing the barge round except in one or two places, the last five miles had to be navigated stern first to the berth. After discharging the barley, we went over to Barrow Haven and Barton upon Humber, where we loaded 170 tons of tiles for London. The rate on that cargo was 5/6d per ton with a gratuity of one guinea to the master. Loading took place from the 19th until the 22nd September and discharging took place at Bermondsey between the 1st and 6th of October.

For our next freight we went down river and up to Upnor where we loaded 142 tons of scrap iron for Goole at a rate of 5/9d per ton. We left Upnor late on October 10th with a southerly wind blowing. By the time we reached Sheerness it had freshened considerably, so instead of fetching up to our anchor, which had been my original intention, we kept going on with the wind from the south which was helping us all the way. Goole was reached on the 12th after a non-stop day and night passage in really good time. From Goole we took the barge to Hull, where a cargo of 130 tons of cotton seed was loaded for discharge at Colchester. The rate was 9/6d per ton, which meant a gross freight of £61.15s.0d. Whilst we were loading, instructions were received from the owner's office telling me to get the barge to Colchester before October 21st or tie up at any port convenient along our route, so as to attend the Admiralty Court in London regarding the collision we had with the tug Crusader back in April.

We rushed away from our loading berth on the 15th. On the way south we encountered a strong south-westerly wind and had a very dirty night off Southwold. It was only with the greatest of difficulty that we held our own off the Sizewell Bank buoy. Nevertheless we managed to get to Harwich and then head for the River Colne, for Colchester. We arrived on the 20th as demanded. That must have been a lucky omen as I won my case, the court finding in our favour and all the blame attributed to the tug.

The next job was to carry a cargo of scrap iron from Upnor to Goole again, loading on the 23rd and 24th October and discharging at our destination on the 29th and 30th. Fresh orders were received to load coal while at Goole for Gweek down in Cornwall. Facing that long trip we signed on a third hand, but as it turned out we only took fifteen days. Loading started as soon as the previous cargo was out, after moving berths on the 30th October. Gweek was reached on 14th November with a cargo of 164 tons, 7 cwts of coal, carried at a rate of 10/6d per ton. We earned £86.5s.8d. together with a gratuity of two guineas to the master, less expenses that totalled £13.14s.7d. From Gweek we went to Pentewan again to load another cargo of grit, part of which was consigned to Gravesend and part to the New Hythe Paper Mills. We loaded on the 25th November and arrived at Gravesend on the 8th December. The New Hythe part of the cargo was discharged on the 10th. It was my wife's birthday the next day so things had worked out very well for us, Kathleen and I able to have a few hours together with the boys.

Linseed was loaded during the 16th and 17th December at the Surrey Commercial Docks. The consignment was for Southampton where we arrived on the 26th, having spent Christmas at sea. Well, 'Better the day, better the deed' as the saying goes. The freight quoted was 8/- per ton on the cargo of 145 tons, which meant the gross earnings amounted to £58. To be set against that were the expenses amounting to £7.5s.1d. The end of the year saw us ordered to Newport, Isle of Wight to load sugar beet.

As the year was closing I sat down in my oil lit cabin writing a letter home to my family. With little else to do I then had time to reflect on what the engine had done for us in the previous six months trading. We had done twelve freights. Taking account of the total distance covered between the Humber in the north, to Falmouth in the south-west, through all weathers and tides, I thought we had done pretty well.

New Years Day 1931 started with us loading sugar beet, 99 tons of it for Harwich. The freight was not very much at only 6/6d per ton. Loading was finished on the 2nd January, and for that time of year we had a remarkably good run up Channel reaching Harwich in something like two days. We hung the side lights up at the Owers lightship and sailed up through the Downs in the dark. To make a short cut to Harwich, as the wind was a little south of east, I steered a course outside the Kentish Knock and we were at our destination by the 4th.

The following day we finished discharging and went up to Ipswich where sugar was loaded for Wandsworth Creek, London. The freight rate which the barge earned for the cargo was 4/3d per ton, well below the Union Blue Book rate of 5/-. More annoying still was that I discovered the charterers had in fact been paid 4/9d per ton. Not being able to do anything about it, I just decided to avoid carrying that type of cargo at those rates again. As it happened we were unable to discharge at Wandsworth until January 15th. Once empty, however, we lost no time in getting down to Strood were 100 tons of cattle food was loaded at the oil mills for the Isle of Wight. Late in the evening of the 16th we put to sea and arrived at our Newport berth on the 22nd, after passing up through the railway bridge to discharge in the centre of the town.

Between January 25th and 26th we loaded a cargo of wooden bobbins at Littlehampton for Grimsby and Hull. We arrived at Grimsby on the 30th and finished discharging at Hull on February 3rd. Bearing in mind that it was by then mid-winter with gales and heavy seas to combat, our passage of four days in a barge such as the *Scone* was most satisfactory.

It was then up to Goole, where coal was loaded for Sandwich. The discharging berth was up through the Sandwich toll bridge where we arrived on 17th February, my second son's first birthday. After discharging we left Sandwich on the 19th for a trip across the Channel to Calais where we were to

Strood Oil Mills building is on the right. These wharves were visited by hundreds of sailing barges in their heyday. One of Cranfield Bros. craft lays beyond the Oil Mills, probably discharging flour from Ipswich. Just off the mill was the Oil Mill buoy where barges would wait for a berth at the mills or for orders. The Oil Mill buoy continued to be known as such for some 30 years after the mills were closed and demolished. (Bob Childs collection)

load pulp for New Hythe Paper Mills. By the early days of March we had returned from the French port and were discharging the woodpulp at Aylesford. We came down river to Strood and loaded for Ipswich 85 tons of meal cake, later returning light to Strood to load 50 tons of oil cake for Dunkirk. We arrived at Dunkirk on the 16th and discharged the following day. Gluten meal for Bow Creek, London, was our return freight which was discharged on the 21st.

From London we carried 111 tons of linseed to the Strood Oil Mill. Next at long last a cargo to fill our holds, for we were bound down Channel with 170 tons of cement to Weymouth. The cement was loaded at the Crown Quarry Cement Works on the 26th March and Weymouth was reached on the 7th April. The freight rate on that cargo was 8/- per ton. The expenses worked out at £16.4s.9d. leaving a net sum earned by the barge of £52. That also had to cover our labour for discharging the cargo ourselves over a period of three days.

From Weymouth we went to Littlehampton for our next cargo which was to be bobbins again for Hull and Grimsby. On the 17th April we were discharging at Grimsby and the following day at Hull. Oil cake compounds for Kings Lynn at the rate of 5/- per ton next filled our holds as we carried a total of

Robinson's Coal Wharf in Littlehampton harbour with the tops'l schooner Adela alongside and the mulie *Viking* of Rochester next out. Outside her is a smart, white painted launch with clipper bow and counter stern, wearing a large ensign.
(Tony Farnham collection)

165 tons from the Old Harbour on the River Hull. We completed loading by the 21st April and arrived at Kings Lynn on the evening of the 23rd. Discharging lasted just over a couple of days, after which we returned to Hull to load a second similar cargo for Lynn.

These were routine voyages in what was a routine way of life. I suppose we were just about earning a living from the freights carried. Such was the commitment to that way of life, many like myself who manned these craft had virtually no time off for themselves.

On the last freight to Kings Lynn the barge had only earned £41.6s.6d. against which were set our expenses. These were made up mostly of costly

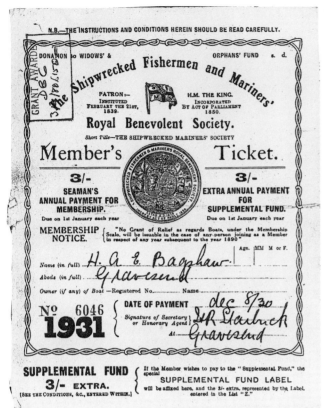

'Harry's' 1931 Member's Ticket for The Shipwrecked Fishermen and Mariners' Royal Benevolent Society, issued by the local Honorary Agent, D & R Starbuck, the Gravesend ship's chandlers. (Albert Bagshaw collection)

harbour dues and compulsory pilotage which amounted to nearly a quarter of the freight rate, some £9.11s.0d. From Kings Lynn, we went light to Gillingham where we loaded cement for Poole. We arrived in the Medway on the 9th May. Loading was completed by the 11th and we were at Poole on the 18th, taking two days to empty our ship. There being no cargoes in the offing at that time from the south coast ports, we had to return to London light for nothing. We couldn't afford too many of those 'labour of love' runs!

Once again we made a good passage back, for Erith was reached on the 30th May. We then loaded a cargo of 153 tons of asbestos slates for Plymouth. Loading took place during the first two days of June and Plymouth saw us tied alongside by the 6th.

Luck was with us that time for we managed to pick up a cargo of barley and maize for Gweek and Truro. The loading occupied the best part of three days and we arrived at Gweek on the 14th June and Truro on the 16th. From Truro we proceeded to Porthoustock inside the Manacle Rocks where we loaded stone for Newhaven at 4/- per ton. After that episode I refused to trade down Channel to the West Coast, for the freight rates offered were insufficient to cover our costs, and there were rarely any return cargoes. We reached Newhaven on the 21st June and discharged the following day, returning light once again for Gillingham to load cement for Hull. We loaded on the 27th June and arrived at Hull on the 29th. That trip turned out to be the start of regular North Sea trading on the East Coast which was to last for several years.

After discharging the cement we went on to Goole to load coal for Gravesend at 6/- per ton. I discovered that our little engine was not really powerful enough to help us on the Goole river and began to think of something bigger. Our coal cargo was destined for discharge at a new quay at Gravesend, known as Bannister's Wharf. It was a destination which our laden barge could only reach with the extra water of spring tides. That meant that although we loaded on the 2nd and 3rd July and arrived at Gravesend on the 6th, we did not finally discharge until the 14th, using our own gear to heave out the cargo in baskets.

Determined to improve the slow unloading, I purchased a motor winch to speed things up. A quayside crane could not be used owing to the narrow frontage of the wharf.

Our next run was from London to Wisbech with barley. Wisbech was a new port to me, but having loaded on the 15th we arrived safely on the 23rd, the freight rate being 4/6d per ton. The barge earned £34.2s.2d. gross, against which were set the expenses of £9.9s.0d. leaving a very poor return again for

The SS Kemmendine makes no smoke, whilst the ship beyond is less kind to those downwind. (National Maritime Museum, Greenwich, London)

the miles sailed and the work involved. From Wisbech we went to Boston from where we carried a cargo back to Gravesend, before going on to Millwall for a new freight, 110 drums of black grease and 70 barrels of acid oil for Hull. As that was only a part cargo, we also took on 80 tons of oil cake for Yarmouth from the SS Kemmendine at Tilbury on our way down river. Although the Blue Book freight rate for that Oil Cake was 8/- per ton we had to accept 5/- and I might add we were lucky to get even that. We arrived at Great Yarmouth on the 13th August and discharged the following day, and were at Hull with the remainder of the cargo on the 15th. After discharging at Hull we loaded 128 tons of caster meal for Strood. That freight was fixed at 7/- per ton, which meant the barge earned £44.16.0. for the trip, against which expenses were only £3.17s.0d. Loading took place on the 18th August and we discharged at Strood Oil Mill between the 23rd and 25th August.

With our firm's yard nearby our thoughts turned to the needs of the barge. It was twelve months since we last undertook any of the routine maintenance we could not accomplish whilst the *Scone* was being worked. Attention was to be given to the gear for wear and chafe. The vang falls, main sheet and davit falls were replaced. The main shroud lanyards were set up tight. Then our all important hatch cloths were repaired. In fact, everything was made safe for the onslaught of winter. Not forgetting our own personal needs, we topped up our water tanks full whenever we were alongside. We also took the opportunity to top up the food in our lockers, coal, wood, paraffin for lamps and primus stoves together with a little meths, and fuel oil for the engine.

From Strood we returned light to Goole for a further cargo of coal for Gravesend. We were back in the Thames and had finished discharging by September 5th. Then it was round to the Medway to load 170 tons of cement for Hull. That voyage was completed on the 16th and earned us £46.15s.0d. with expenses totalling £7.15s.1d.

Our journeys continued with the familiar pattern of passages to Goole for Gravesend with coal. Having found that charter myself, and made a go of it, it became very much part of our way of life. It had the advantage for me of providing regular work and giving me frequent visits home. The petrol driven water cooled motor winch was a boon to that trade and proved a good investment. We had it bolted down to the deck over on the starboard side of the mastcase.

From the Thames we carried about 100 tons of middlings to Great Yarmouth, the freight at 6/- per ton, and were unloaded by October 10th. From Yarmouth we carried sand to Hull, but the rate was very poor for we only secured 3/6d per ton. By October 26th we had discharged and a few days later we were southbound again with our regular coal for Gravesend, which was clear of our holds by November 4th. As soon as she lifted on the flood we were off up to London to load 854 quarters of barley for Ipswich. Although we made Ipswich on the 9th we had to lay waiting until the 13th before we discharged. That was the usual way of things at Ipswich. They always took the full number of lay days allowed[1] so we regularly gave them free storage.

[1] *Lay days were those unpaid within the time allowed for unloading when the barge was not worked, as distinct from demurrage days which were those beyond the time allowed, and for which payment was made whether the barge was unloading or idle.*

We went on to Great Yarmouth to load sand for Hull again. The secretary of the owner's company wanted us to go for a cargo offered at Dunkirk. Coming at that time, I am afraid his wishes led to an argument between us, for having secured a regular run, we wanted to keep it. Needless to say we were left alone after I threatened to quit again! Having just received the last cart load of sand down the hatches on the 18th November we began covering up. It was while we were battening down I became aware that the wind had shifted round to the south-east. That could only mean difficulty for us getting out of Great Yarmouth harbour. However our motor, in spite of its diminutive size, helped by our sails, fulfilled its role successfully. Our cargo of 188 tons of sand brought the barge down very low in the water. That made things very uncomfortable for us as we met the seas after clearing the piers. Once through the Cockle the barge soon steadied herself down for the night's northbound run.

After the sand was unloaded we took on our coal at Goole for Gravesend. We had to bring up to our anchor when we arrived off the Ship and Lobster on November 27th to await the spring tides. It was not until the 30th were we able to reach the berth. After discharging during the first two days of December we went to Gillingham to load cement for Hull, reaching the Humber on the 14th and discharging the following day. By way of a change we took on a cargo of oil cake for Wells. Loading from the 19th to the 21st we arrived at Wells on the 23rd and completed our discharging by the 30th, spending Christmas in the busy Norfolk port.

Thus we completed our voyaging of 1931. For us it was a milestone for we had clocked up twelve months trading with the assistance of our engine. We had carried 35 cargoes, against the 24 which we would have typically carried under sail alone.

The engine allowed us to trade more efficiently and get about quicker, but we worked just as hard. No eight to five for me and the hardworking mate, no weekends off, just us two and the *Scone* keeping keeping hard at it when the weather let us and the work was available.

CHAPTER 14

Narrow Escape

"Perhaps one could say the most dreaded enemy afloat, something we just hoped would never happen to us"

The New Year of 1932 opened with us loading our familiar freight of coal from Goole, arriving at Gravesend on January 9th. Discharging was finished by the 13th and that was followed by a short river freight of 172 tons 14 cwts of bagged wheat from the Victoria Dock to Poplar Dock. We had to load ourselves from a bagging machine, and believe me the four bushel sacks weighing 240 lbs wanted some stowing as they came down the shute. It was a quick freight however, for after docking on the 15th January we loaded and discharged the following day, earning us £38.17s.0d. less £2.1s.3d.

Our next assignment was a mixed cargo, firstly 100 drums of oil for Hull and to top up 1,010 bags of ground nuts for the Strood Oil Mill. After discharging the nuts at Strood we loaded 239 barrels of black grease for Hull on the 22nd January, arriving there four days later and discharging on the 27th.

We sailed for Goole the next day and were back at Gravesend and our coal unloaded by February 8th. We were given orders to go up to the West India Dock to load 170 tons of linseed for Ipswich. Arriving on the 10th, we loaded between the 11th and 13th and lost no time in arriving at Ipswich on the 15th February, gaining one days demurrage because we were not discharged until the 17th.

Returning light to London we loaded 165 tons of wheat at the Victoria Dock silos for Grimsby at 5/- per ton. That turned out to be something of a protracted voyage, for bad weather forced us to take shelter in Harwich and Great Yarmouth. Our little 30 horse power engine was proving not to be man enough to help us on occasions like those. And it was certainly lacking sufficient power to rely on when encountering the strong tides of the Humber. That in turn led me to ask our owners for a larger power unit.

We continued to follow our pattern of routine goings to Goole. After discharging at Grimsby on the 14th March our coal was loaded and Gravesend saw us again on the 21st March. Emptied the following day, we then went up to the Millwall Dock where 650 quarters of bagged wheat was loaded at 1/6d a quarter, which meant the barge earned £48.15s.0d. before expenses.

That consignment was for new waters, through Great Yarmouth for Beccles on the Norfolk Broads. Up the Breydon Water we went, though once again the engine proved to be hardly powerful enough for the job on hand; somehow we managed it without getting into any trouble. We found our way inland via the Haddiscoe Cut, arriving at Beccles on the 29th March. We were empty on April 2nd, then retraced our way, weaving amongst the reeded banks of the waterway to the sea, whence we turned north for Goole. Loading our usual coal cargo on the 9th, we made our way back to Gravesend, clearing our holds by the 20th.

From Gravesend, we went round to Strood where we loaded 219 barrels of oil and 108 drums of black grease. Then we had to double back up to London to top up with a similar cargo from Millwall. In all the loading time, including shifting

'Harry' returns aft as *Scone* makes her way under power, mate at the wheel, after unloading at Beccles on the Norfolk broads.
(Albert Bagshaw collection)

from Strood to London, took from the 26th to the 29th April, but we were at Hull on May 3rd and were finished in a couple of days.

Although the cargo only totalled 100 tons, our barge earned £59 with expenses coming to £9. We were under the coal hoist at Goole by the 6th and were heading south the same day. The cycle of events virtually repeated themselves, for after discharging at Gravesend we went up to Millwall again to load grease from Armstrong's Wharf. We loaded a part cargo on the 13th May then made our way round to Strood to complete the freight. We left the Medway in our wake on the 18th and arrived at Hull by the 22nd. For that run the gross earnings were only £40, with expenses coming to £7.3s.8d. Our own oil bill averaged out at about thirty shillings per week, and as that cost was set against the crew's share of the freight money, we naturally sailed as much as possible.

On that occasion we went from Hull up to Selby to load oil cake for Kings Lynn. Selby was well above Goole and we had only our little engine to get us there. We arrived at the Lynn Roads on Saturday night, and naturally we were looking forward to dropping the anchor and having a good night in. However, that was not to be as the pilot cutter came alongside and asked me to go in that night. I had a number of objections to their request. For one thing I knew perfectly well that the berth on the foreshore which we should occupy would have to be dug out before it would be ready for us. Meanwhile we would have to lay at the buoys during the ebb tide that night, and by so doing it would entail us paying the pilot for mooring us, and again next day for moving us to our berth.

Finally after much argument, and against my better judgement, we went in and tied up to the buoys. Sure enough my prediction proved correct! Having obliged the pilot by moving in they charged us heavily for the subsequent shifting and re-mooring. That led to a heated argument between myself and the pilot master, but to no avail.

We moved to nearby Boston to get our coal for Gravesend. We finished loading on June 10th and set about our voyage south. After we had put a few miles behind us we anchored for the tide, just off Dunwich on the Suffolk Coast. We had a brew up and a bite to eat. After seeing things were all secure on deck, we went to our respective ends to get our heads down for a well earned rest.

Later I was awoken from a sound sleep by strange banging noises. I shook my head to clear my thoughts, for I had never heard anything like it before. Over the years one got to know the little moans and groans of the barge, like the kicking of the rudder and the slatting of the gear aloft. The banging continued above me on deck, by then accompanied by the dreaded shout "Fire! Fire!" I clambered up the companion ladder through choking smoke to the deck, alarmed and desperate to find out what was going on. The strange banging noise had been from wielding axes. Strangers, fishermen in fact, were chopping at our deck. They were nearby and had seen smoke rising from us and came to investigate. Having boarded they discovered the smoke was coming from the engine room. It was lucky for us they were in the area, for without them the fire would have taken hold and the barge and perhaps ourselves lost. Together, and only then with the greatest difficulty, did we finally get the fire under control and eventually extinguished, having to chop a

Scone on fire, pictured in the Suffolk Chronicle and Mercury in its issue of 17th June 1932. The view that the fire 'was easily extinguished' is not in accord with those at the scene.
(Suffolk Record Office, Ipswich)

SOUTHWOLD.—The spritsail barge Scone which caught fire off Southwold early on Sunday morning. While attempting to reach the harbour, the vessel grounded, but refloated at high tide. The fire was not serious, and was easily extinguished.
(Photo, Leyneek. Beccles.)

good deal of smouldering decking away in the process. By that time the Lowestoft lifeboat was standing by us.

Our investigation as to what had been the cause got nowhere, despite considering every possible cause I was never to know for sure what had happened. It was just possible that the few oily rags which I kept on top of the fuel tank caught fire, although the tank was high up and long way clear of the engine over on the port side. What caused them to catch, if they did, when I was a non smoker and the engine room was my exclusive domain, I just did not know.

BARGE ON FIRE NEAR SOUTHWOLD.

RESCUE WORK BY LOWESTOFT LIFEBOAT.

Holiday-makers at Southwold were provided with a thrill on Sunday morning, when the Rochester barge Scone caught fire while she was hove-to only 100 yards from the shore. The vessel, which was bound from Boston to Gravesend with a cargo of coal, is equipped with both motor and sails, and the fire started in her engine room, being first discovered apparently by local 'longshore fishermen, who were fishing near by. They gave the alarm to the crew of the barge, and a salvage party went aboard.

As the Southwold lifeboat is undergoing an overhaul, a call was sent to the Lowestoft lifeboat about 11 o'clock, and the Agnes Cross promptly put to sea, her rapid progress down the coast being followed with interest by the crowds on the Lowestoft South beach.

An attempt to reach Southwold Harbour had been made, but the vessel grounded, owing to low tide, just outside. The fire did not prove serious, and was without great difficulty extinguished. The barge has put into the harbour for repairs.

The Suffolk Chronicle and Mercury report of the fire describes this near disaster as 'a thrill' for holiday-makers. (Suffolk Record Office, Ipswich)

We said goodbye to our fishermen friends and got underway. I decided we should put in to the nearest harbour, so we headed for Southwold, accompanied by the lifeboat. At first we found too little water over the bar at the harbour entrance and grounded there. We waited for more of the flood tide, and once inside we got the damage examined and I made my report to the owners. In my closing remarks to the office, I requested them to pass on the news to my home. Later my wife told me how she had received news from the owners by phone "*Scone* has been involved in a fire at sea but she's alright." Kathleen answered that startling news by saying "Never mind the boat, are the men alright?" She was reassured to be told that we were, but was told of my eye troubles.

I had suffered the effects of the smoke from the fire. My wife came up to join me in Southwold with our elder boy, leaving the youngest with his grandmother for almost a week. It took that time for my eyes to get a bit better after being very red and swollen. In fact my left eye never did recover fully from that awful incident.

As soon as I thought I was well enough, I said goodbye to my wife and sailed for Gravesend, arriving on the 17th June. After discharging we went up the Medway to our Strood barge yard for repairs. That included having practically the whole of the port quarter renewed where the fire, plus the chopping activities had caused such extensive damage. That really seemed to upset the barge as, in spite of all the caulking, it was from that moment in time that our port quarter suffered a number of bad leaks. We never did get it all properly tight as it was before the fire. When we were loaded it became an on-going problem for many years.

The first trip from our repair berth at Strood was to London with 100 tons of scrap iron at 2/3d per ton, less than filling the barge so we could see how she performed in the river. Our next orders were to load linseed for Ipswich; 2,330 bags from Bellamy's Wharf, Rotherhithe. That gave us enough for a sea trial, which was also completed without problems. Arriving at Ipswich on August 20th we finished our discharge five days later. We were then back in our old routine, for after leaving Ipswich we made our way light down to Goole to load more coal for Gravesend.

HOME TRADE SHIP OR
FISHING BOAT.

Certificate of delivery of Half-Yearly Returns.

ISSUED BY
THE BOARD OF TRADE,
in pursuance of
57 & 58 Vict. ch. 60.

I CERTIFY that *A. S. Bagshaw* Owner of the *Scone*
Master
of *Rochester* Official Number *127269* has this Day deposited
with me the Agreement, Official Log Book, List of the Crew, and all other Documents required
in accordance with the provisions of the Merchant Shipping Acts, for the Half-Year ended
June 30th 19*32*, and has produced the Certificates of the Master, Mates and
Engineers required by the Merchant Shipping Acts.*

Dated at

14 JUL 1932

Day of 19

Superintendent.

*Delete as necessary

Every six months the *Scone's* log and crew list were deposited as required by the Merchant Shipping Acts, on this occasion at the Mercantile Marine Office, 2 Manor Road, Chatham. (Albert Bagshaw collection)

We left the coal berth and made our way up river, having to lower our mast down to go up through the bridges to Blackfriars where we loaded 101 tons 15 cwts of paper for the French port of Gravelines, discharging on the 14th September. Calais was our next port of call to load 130 tons of woodpulp for the New Hythe Paper Mill.

Back on the Medway we had to lower our gear ready to pass under Rochester Bridge on the early flood. At our destination it had to be hoisted up again so we could get to our hatches to discharge. From the paper mill the process had to be repeated again for us to pass under the bridge as we made our way light out of the Medway up to Armstrong's Wharf at Millwall, where we collected 169 drums of oil for Stoneferry at Hull. The freight rate for that cargo was 11/- per ton, which gave the barge about £40.10s.0d. before expenses. By the last day of September we had completed loading and had started our run run down the coast. We reached Hull on the 6th October and a few days later we were loading coal at Goole for Gravesend again.

Those coal runs went like clockwork, for we were off Gravesend at the Ship and Lobster by the 15th. Our freight that time was 162 tons at 5/9d. Discharging took place between the 17th and 18th of October. Once more we left the berth to go up river to load wheat in the Surrey Dock, destined for Beccles. After loading our 650 quarters between the 19th and 20th from the SS Beaverburn we made our way to Great Yarmouth. As before, we went up the Breydon Water and through the Haddiscoe Cut, having to wait for trains to pass at two railway swing bridges. That freight had a rate of 1/3d per quarter, which brought in a gross sum of £40.2s.6d. out of which £7.19s.1d. was paid for expenses.

We left Beccles for Orford Haven to load shingle for Southend jetty. 165 tons came aboard at a rate of 3/- per ton, and by the 4th November we were empty and awaiting the next assignment. From Southend we sailed up to Armstrong's where we loaded a mixed oil cargo of 100 tons for Hull. We were at Hull by the 10th and our holds clear by the following day. Thereafter came another cargo of coal for Gravesend, where we returned on November 18th.

During that time freights in general were still becoming more and more scarce. That was borne out by the fact that our next job was shuttling for small parcels of cargo between London docks and wharves and the Medway, while awaiting a longer run.

The SS City of Roubaix which had brought linseed to Tilbury Docks. (National Maritime Museum, Greenwich, London)

From Strood we picked up 17 tons of oil in barrels for Hull. Then we sailed out of the Medway up to Tilbury Docks to load out of the SS City of Roubaix 1200 bags of linseed for Kings Lynn. Our first destination was reached by the 5th December. Once our linseed was discharged we found there was a strong easterly wind blowing, making it unwise for us to continue our voyage to Hull.

I knew only too well that if the weather came bad we might have had days or weeks to await a fair slant in the wind to keep us busy. I arranged to take some coal down to a ship lying in Lynn Roads awaiting spring tides to enter port. The Captain of the ship then asked me if we could take off some deck cargo, which we did after agreeing the rate. That extra work was paid the same as the coal we delivered to him, £5 per day. When the barge was loaded we found there were no pilots available, so undaunted we went up to Kings Lynn and into the dock unassisted. As soon as we had arrived I was faced with a belligerent demand for pilotage fees. I refused point blank to pay, pointing out that no pilots had been available, and furthermore we had not left the port.

The final outcome of that little argument was that for the future we arranged only to pay the pilotage fee for entry and clearance from the port once, and that was irrespective of how many times we moved berths or loaded and discharged, and whether we took on a pilot or not.

That arrangement continued satisfactorily for many subsequent occasions when I traded to the port, for we never worried the pilots again and they in turn never interfered with us. We would come and go as we pleased. As for shifting moorings and berths, I dare say that by then we knew more about the Lynn Channel than some of the pilots did!

Once clear of Lynn, we made our way to Hull with the oil from Strood. From Hull we went to Goole for our usual coal cargo, which we brought back to the Thames arriving at Gravesend on Christmas Eve.

Our working year of 1932 finished, as far as we were concerned, when the last piece of coal left the hold at Gravesend on December 29th. Although the year had seen us lying on the shipyard whilst rebuilding our quarter for six weeks following the fire damage, we had nevertheless completed thirty voyages.

CHAPTER 15

A Night to Remember

"I decided to snug the barge down in preparation for a bad night ahead."

The New Year of 1933 was only two days old when we arrived at Strood from Gravesend. As soon as we were alongside we started loading 85 tons of cattle feed for Ipswich. The weather, surprisingly for that time of year, gave us no trouble and we reached our destination by the 4th. We left again on the 6th light for Boston where we arrived on the 8th January. There we loaded 154 tons of coal for Gravesend. By the 13th we had returned to my home town where, with the aid of our little motor winch, discharging took place over the 14th and 15th.

My purchasing of that little machine for that particular job was certainly arousing the owner's interest. Realising its worth in giving us a speedy turn around, they paid me for it, and installed another one in the barge *Alderman*.

Clearing the coal berth, our passage was to Grays where we found a cargo of 'pickled timber' creosote soaked logs awaited us. In my opinion the acceptance of that cargo for a craft such as ours was absurd. Even a layman would quickly realise that the barge, being built of wood, would soak up the creosote. Future general cargoes would absorb the penetrating reek of the creosote left behind in the lining of the barge's hold and be the subject of claims for damage. However, we were stuck with the freight so, to reduce the risk of problems, I obtained some bundles of battens which we used to cover the ceiling. As an extra precaution we spread a liberal covering of sawdust throughout the holds.

With all the preparation completed we loaded the 64 timber baulks for Hull. We were required to make two more trips to obtain a reasonable size freight before we could put to sea. From Grays we went up to Armstrong's Wharf at Millwall to pick up some 42 drums of grease, and then came down river again to turn our way up the Medway to Strood to complete our freight with 15 drums of oil. After covering up and battening down for the third time we finally put to sea, arriving at Hull on February 2nd.

From Hull we went once again to Goole, where 161 tons of coal was loaded which we took back to Gravesend at 5/9d per ton. Leaving Gravesend on the 13th we went up to the Erith Oil Works, where we loaded 1,120 bags of soya meal for Ipswich. Once unloaded we sailed light to Boston again, arriving on February 24th. 158 tons of coal was our freight for Gravesend, where we arrived on March 8th.

Our next orders were almost a carbon copy of the last, another freight of creosoted timber for Hull. In spite of my objections regarding carrying such cargoes, we were again obliged to accept the freight. We took the same precautions as before, and as previously we made further calls to complete a full cargo. First to come aboard was 28 drums of acid oil and 74 drums of black grease from Millwall. From Silvertown and Erith, we took on another 67 drums of black grease and 31 drums of oil. With that additional mixed cargo, all in steel drums, we set out for Hull.

We discharged there between the 21st and 22nd March and we loaded 5046 drums of soft soap for Chatham Dockyard for the lump sum of £40. That cargo

was discharged at Chatham over the 27th and 28th. Sailing light we left the Medway bound for Orford Haven. Our freight was to be 140 tons of shingle for Grimsby at 3/6d per ton. By the 7th April we were in the Humber and discharged. That was followed by a short light run to Goole from where we loaded 157 tons of coal for Gravesend. By the 14th April we were in the Victoria Docks, London, loading barley.

We were two days loading 2,055 bags from the SS Asburtor for Wisbech, the only port in Cambridgeshire, where we delivered by the 24th April. The gross on that freight came to just £35 against which expenses amounting to £9.8s.11d. had to be paid. That was a lot of expenses and at that time I found all the Wash ports were expensive to the barge, various dues, compulsory pilotage fees and so on. From Wisbech we proceeded to Boston where we loaded more coal for Gravesend. By the 11th May we were on our way up river to Armstrong's Wharf again. Once more we were to load for our run down to Hull with oil and black grease. After discharging we loaded oil cake and compounds at different mills in the Old Harbour and set sail for Kings Lynn where we arrived on the 27th. Our luck was out that time for we did not get a berth until the 29th, but we had discharged by the 31st and earned £25.2s.9d. after paying expenses of £5.16s.0d.

We left Lynn and sailed light ship to Goole where we loaded 151 tons of coal and sailed south for Gravesend. Discharging completed, we left the berth and went up river to Silvertown and Millwall to load for Hull again with oil and black grease. After unloading ship we found we were due for a change. Instead of the usual return cargo of coal we were to load tiles at Burton Stather on the River Trent for delivery at Norwich.

The cargo was quoted at 12/- per 1,000. Into the holds went 30,000 tiles of one type, followed by 50,000 of another, then 5,000 ridges. The total freight came to £48.12s.0d. The loading took from the 22nd until the 24th June. Norwich greeted us on the 26th but we did not finish until the 30th for we could only move around 50 tons a day and we did not get started straight away. From Norwich we went light, first to Barton on Humber, then to Hull, from where we loaded 90 tons of mixed tiles and 40 tons of poultry feed for London at 6/- per ton. Our loading was four days.

We left on 8th July and on the way south encountered strong southerly winds which forced us to put into Great Yarmouth for shelter for a few days. We were not able to reach the Thames until the night of the 17th. Great Yarmouth dues for our shelter cost us 19/9d. After unloading the poultry feed at Bow Creek and the tiles at Bermondsey we came down light to the Medway to part load, first for Hull from Strood Mills and for the topping up we sailed back up the Thames to Armstrong's at Millwall and John Knight's at Silvertown.

In all, the total freight only came to a paltry 71 tons. With the hatches on for the third time and battened down, we finally sailed for Hull where we arrived on the 29th July and discharged a couple of days later.

Our little engine had worked quite well during that time, but nevertheless I was not completely satisfied as we had found it not powerful enough, just when one really needed that little bit extra. My proposals for a new more powerful engine fell on deaf ears, so we had to soldier on with what we had.

The first week of August saw us bring another cargo of coal from Goole to Gravesend. We arrived with 163 tons on the 5th August but owing to the state of the tides we were not able to be discharged until the 9th. Leaving the berth on the high water we proceeded to Strood again where we loaded a part cargo for Ipswich. Then it was back up to London for some more for Hull. Finally we sailed for Ipswich where we arrived on August 20th, and Hull just under a week later. The return passage was the usual run with coal to Gravesend where we arrived on September 4th.

A new cargo awaited us in our next orders. From the coal berth we went up river to the Erith Loam Wharf to load 125 tons of loam at 3/6d per ton. With the barge only part full we moved berths to top up with 82 drums of black grease and 12 drums of acid oil from Armstrong's at Millwall. Once our discharge at Hull was completed on the 20th September we sailed light over to Barton on Humber where we started to load tiles again for Norwich.

When we were on these types of berth we took the opportunity to get over the side and give the barge a tar round, weather permitting. In those summer

Kathleen with Albert aboard *Scone*.
(Albert Bagshaw collection)

months I started having my family aboard for a few days at a time. The boys were getting older and they especially enjoyed the inland passages with so much to look at compared with time spent at sea. It was to become a regular holiday treat in their younger years. Once we were berthed alongside they would have the boat down and play all day, with the exception of meal times. At the end of the day they didn't want any telling to go to bed. It was good that we were together as a family. After all, with *Scone's* comings and goings I hardly ever got home to spend time with them.

We discharged our 168 tons of tiles over three days, completing by September 28th. Returning light to the Humber we loaded more coal for Gravesend. We were always busy in spite of the great trade depression which had brought such hardship to our country. The cotchel freights were a consequence of the slump, small parcels of different cargoes collected from various places and sometimes for delivery to more than one destination. Although the work was hard, and we were about long and odd hours doing our job, the time passed very quickly in those days. It was already the first week of October when we had finished discharging our coal and by way of a change we were ordered up to the Surrey Docks where we loaded 28 standards of timber, some of it above deck, between the 9th and 11th. It was consigned to Sandwich where we arrived on October 13th.

Returning to London empty, we loaded from different wharves taking on 17 tons of acid oil, 34 tons of black grease, 25 tons of bonemeal and 35 tons of flour. Owing to the rainy weather, loading lasted from the 19th until the 28th, during which time we had the extra work of covering up for every shower. The freight grossed £55.7s.0d. before expenses. Finally we sailed and made our passage to Hull by the 7th November, and one week later we were back again at Gravesend ready to discharge our 160 tons of coal.

The next cargo was awaiting us in Deptford Creek, which meant we would be passing up through the railway bridge. It only opened up to give a clearance of eight feet. We therefore had to lower our gear most of the way to get our mast through the bridge hole, then heave it back upright again afterwards. In time, largely thanks to my own efforts, we became quite a regular trader there.

We had on that occasion been fixed to load scrap iron for Hull at 7/- per ton, the barge paying for the discharge at the destination, which came to £6.14s.9d. I had always objected to that arrangement, for one had to pay commission on that part of the freight. I talked the matter over with the manager of the works concerned and finally came to an agreement whereby the barge would take in a cargo from any of their scrap wharves in London at a rate of 5/6d per ton. The arrangement was to apply to large or small quantities with free in and out, known as F.I.O. That meant that the freight rate only had to pay for the carriage of the cargo; all other charges such as loading and discharging were borne by the shipper, not the barge. I wrote to our owners informing them of the full details of the arrangement that I had made. Naturally enough I suppose, the Director of the firm was furious with me for arranging such a scheme, "To lower the freight charge indeed!" However, after an explanation of the arrangement, which I felt should hardly have been necessary, he realised that much of the higher freight rate under the old arrangement was so much dead money, of no use to them or us, as it was paid away for the extra expenses incurred.

That arrangement secured the business for our company for many years; in fact right up to the start of the World War II, and a very profitable arrangement it was. Moreover, from our point of view it fitted in well with other part cargoes, which we were by then regularly carrying north. Things were picking up so I took on another lad as third hand.

The first cargo carried under the arrangement was loaded between the 20th and 24th November. We arrived at Hull on the 30th. After discharging by the 2nd December we made our usual short run to Goole for our Gravesend bound cargo of coal.

Leaving Goole on the 6th, we sailed down the Humber and out of the river past Spurn Head, our heading south'ard with the wind moderate from the north-east. We were making a good passage until daylight on the 7th, when the wind went into the south-east just as we were off Cromer and the swell began to make up. Knowing there to be no safe haven for us on that part of the coast and with the wind then blowing hard, I decided to run back to the Boston Clayhole, up the Wash for shelter. We anchored in the late afternoon just before dark. While it was a good anchorage, we found it a different story on top of the tide. We had to cover up the fo'c'sle hatch and give the barge more anchor chain to hold her. Even so she was still snatching. Next morning when the tide was low we hove in some 30 fathoms and put our six inch bass rope on to act as a spring. Handling our anchor had been made a lot easier since I had rigged up a system to use our motor winch by the mastcase to recover our cable. Slacking away again to give her full sixty fathoms of chain, we laid much quieter, though being laden we were still swept by some of the worst seas.

After a couple of days the wind eased off so we put to sea again and got up to the Wold. There was still a heavy swell running and to add to our difficulties fog descended upon us, making the chance of finding the Cockle Channel very

Scone receives her cargo of coal, and her share of the coal dust that accompanies it, from the chute at Goole. 'Trains' of 8-10 tub boats known as 'Tom Puddings' or 'pans' and towed by tugs, arrive by canal from the Yorkshire collieries. Each coal laden pan is raised in the hoist before being tipped into the chute, prior to being lowered back into the water to make way for the next. From the chute the coal enters the holds via the hopper, giving more accuracy and control to what would otherwise be a rather haphazard procedure. The hopper is suspended by a four part cable below from the angled 'crane' jib. A narrow plank can be seen spanning the gap between quay and barge, for *Scone* has to lay off the wharf, the chute mechanism designed for loading steam colliers of greater beam. Empty pans in the foreground await a tow back to the pit. This coal hoist has been restored and preserved. (Albert Bagshaw collection)

doubtful. We had a light easterly breeze giving us fair steerage and we strained our ears hoping to hear the lightship or the North Cockle bell buoy through the fog, but to no avail. I asked the mate to take a sounding with the lead to see how much water we'd got, knowing that the sands were nearby with their steep-to sides. I decided to anchor when the mate called out eight fathoms, it being around noon. There was a ground swell running and I was far from happy; however, we let go and had a meal, with a watch being kept and the fog bell rung every minute or so.

By three in the afternoon the fog lifted sufficiently for us to up anchor and we passed the Cockle lightship. I did not like the look of the weather over towards the east of us with its heavy clouds and the wind freshening again from that quarter. I decided to let go the head of the topsail, putting the mate at the wheel to pass through Yarmouth Roads. I asked the third hand to go to the fo'c'sle to light the navigation lights and put out the fire. The lights were brought on deck and hung on our light boards by the main rigging. Once that was done we battened down the fo'c'sle hatch against the seas that were beginning to break over our bows. Before making the hatch secure the third hand moved his gear aft to share our very limited space until conditions improved. Darkness descended unusually quickly that December night, and made the shore lights shine eerily bright.

As we entered the Stanford Channel off Lowestoft I had a good deal of anxious thoughts. The wind was increasing, still from the east, and we were on a lee shore with the ebb tide setting against us and we were making heavy weather of it. I decided to snug the barge down in preparation for a bad night ahead. The ventilators and the stove chimneys were securely plugged and covered and the pump was got ready for use. I took the hammer and went round all the hatch wedges making sure they were tight home. Then my thoughts turned back to the sails. We picked up half the mainsail and made all the brails well fast. The end of the mainsheet was stowed in the mooring ropes on the hatches and the whole lot well lashed down.

The topsail sheet was let go and both clewlines well secured. I then decided to reef the foresail, a thing not normally done in a sailing barge. I had arranged with our sailmaker to have our foresail made with extra eyelet holes for that very purpose. I lowered the sail down and cut it free from the lower hanks on the forestay. Then the short wire sheet was removed from around the fore horse before I rolled and reefed the sail, putting on a longer sheet wire and bowline rope. A shackle replaced the lower hank to allow the tack to be replaced. The sail was then reset with about six feet of canvas rolled up from the foot, reducing the area by about one third.

Making my way aft to take the wheel, feeling satisfied with those precautions, I told the mate to go below, saying I would knock the deck if he was wanted. Although the *Scone* was deep in the water she seemed to be riding comfortably. The Southwold light could be seen flashing its warning, and beyond it Orfordness. In quick time we made the Shipwash and began running up inside the Shipwash Sand. The mate popped his head out of the cabin hatchway and asked me whether I intended going into Harwich. I told him that we would be safer in deeper water, and most probably there would be a lot of broken water about if we tried. It was far too risky in such weather. We decided to carry on up inside the Sunk lightship; that far we had not been shipping much really heavy water over us, only spray, spindrift, and the occasional wave crest.

That was soon to change, for as we left the shelter of the Shipwash in order to get under the lee of the Barrow Sands we found there were huge seas running. At that point of our passage we had the whole of the North Sea open to us with the full might of the gale blowing from the East. Once there, we just had to take it.

The Barrow Deep light vessel was under our lee and the light of the Gunfleet pierced the night as we battled south. Suddenly, as the Gunfleet flashed red, the barge plunged into a trough of sea and was buried beneath a wall of water. The *Scone* just disappeared until I was standing up to my waist in water. In front of me I could see nothing but our sails and spars. As I clung to the steering wheel trying to keep my balance my thoughts raced. Had I made an error of judgement and were we all done for? In moments which seemed like hours *Scone* rose up and freed herself from the clutches of the sea. What a relief it was to see the familiar outline of the deck fittings re-appearing from beneath the water.

I sighted the Barrow Deep light vessel again and was reassured to find we were still on our course. Although we were still shipping a lot of water there were no more seas like that big one. I knew that if we could hold our course and get up under the lee of the Barrow Sands we should be alright. With that objective in my mind I kept station, fighting to keep the barge under control in those frightening conditions. I had already been at the wheel from the Cockle light and deep inside me I knew that if we could safely reach the Mouse the worst of our problems would be behind us. We slogged onward, clawing our way to safety, each mile behind us improving our chances of coming through relatively unscathed.

By the time the Mouse was abeam I felt bitterly cold, still standing there in my soaking clothes. The mate must have heard my heavy movement on deck as I tried to keep warm, for without any knocking on my part he was ready to take over, allowing me the chance for a change of clothing and to have a much needed rest. And rest I did for some hours in the knowledge that my mate was a competent hand, always content and willing to take on his fair share of the rougher tasks, along with the smooth.

When I came on deck again the barge was already off the Cliffe Cement Works in the Thames. It was still bitterly cold and I relieved the mate at the wheel so he could have a break. Miraculously *Scone* and her cargo had escaped without damage, discharging our coal on the 16th December. Perhaps a combination of skill and good fortune had seen us through. We had a well built and well maintained barge, though in weather such as we had experienced, that on its own was not always enough. It had been a truly dirty night, and not for us alone. The *Glenway* had gone ashore in the Wold, the *Sepoy*

A deep laden barge in a gale. Wild seas cascade aboard to bury *Scone*. Her bow disappears, a wall of water poised to engulf the length of her deck, submerging everything in its path, to be followed by more of the same as she battles on to safety. (Albert Bagshaw collection)

The *Sepoy* ashore at Cromer; the pulling lifeboat launched twice but was unable to rescue the crew of the barge seen clinging to the rigging.
(Photo: P A Vicary, Tony Farnham collection)

had been lost at Cromer, and the Dyke light vessel had turned over, as had the steamer Broomfleet with the loss of all hands.

From Gravesend we went up to Deptford to load scrap iron again, finishing by the 23rd December and arriving at Hull on Boxing Day. For us it was work as normal. That was our 33rd freight of the year. All our loadings and dischargings had been accomplished without serious problems. The holds had been kept clean, the ropes and other gear had been replaced as required and kept in good condition. Whenever we had taken the ground on a hard berth the opportunity had been taken to give the hull a good coat of tar. I suppose it had been a routine twelve months, a routine however, which demanded hard work from two men, and sometimes the additional services of a boy as third hand, a commitment for 365 days a year, with no time off for pleasure or holidays.

With the hull of the *Sepoy* hidden by the raging breakers, Coxwain Blogg put his motor lifeboat onto the deck of the barge and saved the exhausted bargemen, a feat which saw off the last resistance to the change from pulling and sailing lifeboats to fully powered craft.
(Photo: H H Tansley, Tony Farnham collection)

CHAPTER 16

**Extra
Power**
*It was a Kelvin engine, as was the one it replaced, but with double
the power."*

On the first day of 1934 we arrived back at Gravesend with 150 tons of
coal from Goole.

The next freight clearly showed how the arrangement for carriage of scrap
metal to Hull was bearing fruit. Our orders were to load a full cargo of scrap
iron in the Regent's Canal Dock for Hull. Loading was accomplished by the 8th
January, our destination reached by the 15th, and our discharge completed by the
17th. 165 tons of cargo had been carried at a rate of 5/6d per ton F.I.O. It
followed, therefore, that the barge earned £45.7s.6d. gross on that run, and did
not have to pay anything extra away for the loading or discharging. Our total
expenses came to only £6.16s.0d. which left a net revenue of £38.11s.6d. That
proved a better result than could have been achieved under the old arrangement
of receiving 7/6d per ton and greater expenses. Under the previous system the
commission payable was ten shillings, then there was the employment stamps
for the men to load and discharge us together with their wages, which together
would have cost much more than the 2/- per ton reduction I had negotiated.

Before leaving the Yorkshire port we visited the Cato No. 2 and Eagle
Oil Mills. Both were oil cake mills on the Hull River, which flowed south
through the city of Hull into the River Humber. Here we loaded a full cargo of
oil cake for Wells.

We made ready to depart on the 19th January and were quietly
proceeding down through the Old Harbour when, to our surprise and
apprehension, we found that the barge was making water. The cause was not
long in the finding. Our anchor had been slacked away under the forefoot and
we must have touched somewhere, which had forced the fluke of the anchor
through the side of the barge. There was only one thing to do; find a berth,
put the barge aground, wait for the tide to leave us and patch the leak. That
we did, and when we finally reached Wells I had a proper repair carried out.
Fortunately none of our cargo was damaged.

We left Wells on the 25th January and went light to Boston to load coal
for Gravesend. We loaded on the 27th and arrived at Gravesend on the 30th.
From Gravesend we went up river to Armstrong's at Millwall, where we took
on the first part of a split cargo. With oil cake aboard, we sailed round to
Strood on the 9th to top up with 132 barrels of Grease. The weight of these
barrels amounted to just over 31 tons, at a rate of 14/-. That gave the barge an
additional £21.15s.4d. for the passage north. We carried that mixed cargo to
Hull where we arrived on the 13th. On that run we had altogether collected a
£70 freight, against which the expenses were only £7.3s.0d. Our discharging
was finished by the 15th February, after which we carried the usual coal cargo
from Goole to Gravesend.

That coal run earned the barge £44.18s.5d. gross and £38.4s.10d. net.
Discharging was finished on the 22nd which meant the barge had earned more

than £100 after expenses in eleven days. I thought that was very satisfactory indeed, especially when one took into account that it was mid-winter and freights in general were still not plentiful.

The next cargo was to be a short haul, from London to Strood with 2,866 bags of potatoes and 45 bags of oats. We loaded from a Dundee ship and the gross freight only came to £24.6s.10d. As soon as the last bags cleared the hold we began loading for Ipswich with 1,985 bags of compounds at 5/- per ton. From Ipswich we ran light to Boston for coal and were back in Gravesend on the 15th March. Our next orders took us up river to the King George V Dock where we loaded 1,935 bags of wheat for Ipswich. Loading took four days, as did the unloading at Ipswich, where we arrived on the 26th.

Discharging remained a prolonged business at Ipswich, especially so with wheat or maize freights, for although the merchants always demanded a rapid passage, once in port they invariably wanted to use the barge as a floating warehouse, taking out their cargoes in dribs and drabs.

From Ipswich we returned to London light, where we loaded 150 tons of linseed from the SS Wearpool which we carried back to Ipswich. With our cargo unloaded we left on the 10th April to go light down to Boston, where we arrived on the 16th. Returning south to Gravesend with 147 tons of coal, we finished discharging by the 21st April. We left the berth as soon as we had water and headed up river once more to load 29 tons of acid oil and 63 tons of black grease at Armstrong's, both at the rate of 10/- per ton. Then we moved to John Knight's at Silvertown, where a further 25 tons of black grease was added. With 117 tons of freight aboard we sailed to Hull, arriving on the 30th April.

Our next orders were, by way of a change from the usual coal cargo, to load 200 barrels of glucose, at an average of three barrels to the ton. Although that freight was for Maidstone, we discharged it into a lighter at Strood for onward carriage to its destination. In the meanwhile, we took *Scone* down river in the opposite direction, partly loaded with 22 tons of black grease which we had taken from the Strood Oil Mills. For the topping up part, we sailed out of the Medway and up the Thames to Millwall, where another 20 tons of black grease was added as well as 11 tons of acid oil. Finally we called in at the Erith Oil Works for another 25 tons of oil before setting out on our passage to Hull. Discharging in the Humber was completed by the 16th May before moving on to Goole where another coal cargo awaited us. We loaded a consignment of 163 tons which we landed at Gravesend on the 29th.

Then came more of our familiar pattern of work, up river loading general cargoes in small quantities at different wharves and quays. Although it involved a lot of moving about we did pick up £61.11s.6d. gross freight, which was very good for those days. Repetitive runs they may have been, but we hardly saw them like that, for we were then reaping the benefit of our earlier efforts building up a steady trade. There may have been times when the returns were so small from the freights carried that we wondered if it was worth all the effort we had to put into it, but our barge was busier than most and we still made a living wage.

We finally left the Thames on the 4th June northbound, arriving at Hull on the 9th. Here we discharged our mixed cargo by the 12th and loaded oil cake for Wisbech, Sutton Bridge and Kings Lynn. That brought in a poor freight return, for it only amounted to £36.16s.11d. net, expenses having come to £11.15s.7d. However it was a job at a time when jobs were difficult to find.

It was not until the 20th June that we were again ready to receive a cargo, and that took us up to Burton Stather, a little village just inside the River Trent. Between the 23rd and 25th of June we loaded tiles for Norwich, where we arrived on the 27th and finished discharging on July 1st. That brought in a gross freight of £45.18s.0d. against which expenses came to £10.17s.2d.

From Norwich we returned light to Boston, where coal was loaded for Gravesend. We were back in the Thames on the 9th July, five weeks after we had left, and finished our discharging by the 11th, after which we sailed for Strood.

The next few weeks saw us on the yard at Strood. The barge received a general overhaul, the sails were sent to the loft for a few repairs and a new engine was installed. My endeavours in that connection were at long last fulfilled. It was a Kelvin engine, as was the one it replaced, but with double the power. With three cylinders, it was rated at 66 horse power which gave us good manoeuvring ability. At the same time it drove the barge at a reasonable speed without the use of the sails. In fact *Scone* was to be good for a steady 5 knots loaded with calm water, and at times 7 knots light, which was very good for the size of the engine. As our down Channel work looked to be a thing of the past, I decided that our bowsprit should be put

A Kelvin K3 petrol start diesel engine of the type installed aboard *Scone* to replace her earlier Kelvin.
(Kelvin Diesels PLC)

ashore. It is an indisputable fact that the great majority of sailing barges which were subsequently given over to power were engined based on the successful installation in the *Scone*.

Of great importance was the fact that it proved very reliable and efficient. That caused me no little amusement when I reflected how my owners were so adamant in their earlier opinion that such an installation could not possibly pay its way, as motor barges up until then had generally been a financial failure.

We were back in commission ready for further trading at the end of the first week in August. After loading 7 tons of cargo from Armstrong's at Millwall we then went to the Erith Oil Works. There we took on some more to complete our freight for Hull, which was reached on the 16th. That voyage only brought in £41 gross. We then took aboard 150 tons of oil cake for King's Lynn. That freight was at 4/6d per ton, grossing to £33.15s.0d. We arrived at King's

Lynn on the 19th and a couple of days later we were returning to Goole to load 157 tons of coal for Gravesend. From the coal berth we headed up river to load in the Victoria Dock. We took on 137 tons of barley from SS Jutland for Strood. Unloaded, we were back up the Thames to Erith where we took on slates for Southampton and Poole.

Dating from 1928, the SS Jutland was a typical tramp steamer of her day. She is seen here loaded with timber, some of it as deck cargo. (National Maritime Museum, Greenwich, London)

After a pleasant voyage over some of my earlier trading grounds we arrived at Southampton on September 9th, and Poole two days later. That run brought in £64 freight, against which expenses came to £9. Unfortunately there was no cargo in the offing for the return, so it was a labour of love bringing the barge back light to the Thames. No cargo meant no money for either the owners or ourselves.

On September 19th we loaded at Millwall and topped up with cargo taken in from the Erith Oil Works on the 20th, both parcels for Hull. That freight brought in just £32 pounds, but being in the right place we were able to obtain a fixture straight away, to carry 159 tons of coal from Keadby to Gravesend.

Once more our bows pointed up river after discharging as we headed towards the Regent's Canal Dock to load just over 160 tons of scrap iron for Hull. We arrived at Hull early on the evening of the 7th. Next morning we moved alongside ready for an 8am start and discharging was completed by the 9th October.

From Hull we sailed to Burton Stather on the Trent to load another freight of tiles for Norwich. Loading was carried out between the 10th and 11th and Norwich reached on the 13th. Six days later our 75,000 tiles were unloaded, bringing in £52 freight with expenses amounting to £11.

Our next assignment was to carry sugar beet from Wells-next-the-Sea to Selby. The cargo of 136 tons was at a rate of 5/6d. By the 26th October we were loading at Hull from the Cato No 2 and Eagle Mills for King's Lynn, arriving there on the 29th. We returned light to Goole to load 164 tons of coal for Gravesend, where we finally found ourselves on November 5th. Just two days later we were loading a little over 161 tons of wheat from the SS Hindanger up in the Royal Albert Dock for King's Lynn. Arriving at

Lynn on the 14th, discharging took place the day after. We left light for Hull, from where we carried oil cake back to King's Lynn. The cargo of 170 tons was placed under the hatches in differing lots. In the middle of the main hold we placed 50 tons of Cake, aft went another 61 tons and 7 tons in bags. Under the mast case area went 15 tons and in the fore hold went 37 tons in bags. We found that way of loading just trimmed the barge a few inches down by the stern, just what she liked. Our discharging was completed by the 20th November. On returning light to Hull we immediately loaded a second similar cargo from Cato No 1 and Eagle Mills. With 66 tons in bags and 104 tons as cakes put ashore, we returned yet again for the third time, only that time our freight was 157 tons. The rate for all three runs was 4/6d per ton.

Leaving the Wash we went to Keadby, arriving on the 3rd December to load coal for the voyage south to Gravesend. A little over two months since we left, we returned to the London River where our 151 tons was landed on December 11th. We followed that with a run up river to the Surrey Dock to load wheat for King's Lynn,

Three of E.J.&W.Goldsmith's ironpots, the sprittie *Britannic* and the mulie *Success*, with another sprittie astern and two keels, round up under tow on the flood at Keadby on the River Trent.
(Lincolnshire County Council, Gainsborough Library)

which by then was becoming a very familiar place to us. After we had docked in the Surrey we learned that the ship from which we were to have received our cargo had gone into the Victoria Dock instead!

A change of plan was decided. It was agreed we could do a quick run down to Strood with a small cargo. We arrived back and locked into the Victoria Dock on December 16th. Loading started right away from the SS Yearby. The following day we were away again heading for King's Lynn, which we made by the 20th, and for midwinter that was a good run. From King's Lynn we went light to Goole for more Gravesend bound coal. We were back in the Thames coming to anchor off the Ship and Lobster on Christmas Day. It was most welcome timing. The barge was alongside the coal wharf and discharged by the 28th.

That was the last of our voyages during the year of 1934, which had numbered an impressive thirty eight in all, even though we had that stay on the ways at Strood. We left Gravesend and made our way up the misty reaches of the Thames heading for Armstrong's at Millwall, and that's where the last day of the old year found us, ready to load the first cargo of the New Year, 1935.

CHAPTER 17

**Shuttle
Service**   *"... we had to wait our turn, and others had to learn to do the same"*

New Year's Day 1935 found us changing berths. Having taken on our first 87 tons of cargo at Millwall, we moved down to Erith to pick up a further 16 tons. By the end of the second day of the year we had sailed round to Strood and topped up another 8 tons. Then, with a reasonable amount of cargo under the hatches, we sailed for Hull where we arrived on January 10th. The following day saw all that discharged, and the day after saw the holds full again. We had taken a part load in from the Cato No 1 on the 11th, and a further part from the Eagle Mill. It was oil cake for King's Lynn, where we arrived on the 15th January. Returning light to Hull we immediately loaded a second cargo for King's Lynn, some from Cato No 1 and the rest from Cato No 2. With our discharging completed at our destination on the evening of the 21st, we sailed light out into the night on our way to Goole. There we picked up 151 tons of coal for Gravesend, where we arrived on the 25th, and had discharged by the 27th. That meant that during the first twenty-seven days of the New Year we had carried no less than four paying cargoes, totalling some 547 tons.

The last day of the month saw us back up at Millwall loading just over 19 tons of acid oil and 55 tons of black grease, before moving to John Knight's at Silvertown to top up with another 14 tons; a total of 88 tons for the run north to Hull, where we fetched up on February 4th.

Our next orders were to proceed up the river Trent to Keadby to load coal. Although a large quantity of coal was shipped from there, it was not a very good

place to go, having just one chute on the river bank from which to load. When we arrived at Keadby on the 5th, it was to find the berth under the chute already occupied by a vessel awaiting their cargo. It was in fact the old motor coaster Heather Pet which was running under the house flag of F.T.Everard as the Assurity. She was a small wooden hulled craft built by Wills & Packham at Sittingbourne just after World War I.

On coming alongside the jetty to wait our turn to load, I was informed by the Keadby shipbroker that another of Everard's craft was expected under

The wooden hulled motor vessel Assurity was in the Everard fleet until 1960, finally being used in a static role at Greenhithe for training, and renamed Ytirussa. (Ken Garrett Collection)

the chute. He intimated that we would be expected to give up our turn in favour of them and extend our wait. That was not to my liking at all. I pointed out to him my understanding that, providing the cargo was available, it was the custom to load strictly on turn. I acknowledged that the ship in front of me was first under the chute, but as I knew that our cargo was ready, we should accordingly follow next, and that we did. The other vessel had arrived in the meantime and had to wait. It was regrettable, but we had to wait our turn and others had to learn to do the same. Of course at times like that things could get a bit heated, but I was always determined to see justice done.

That was only one of a number of occasions where queue jumping was attempted at loading berths. A similar incident occurred when we were loading on the beach at Orford Haven and a little motor ship, the River Witham, came on. Her master wanted me to come out of turn and wait a couple of days while they loaded, and had the audacity to offer me £5 to do it. My reply was a firm "No thank you!" for we also had to run to time as he did. The River Witham went away.

It was February 9th when we arrived at Gravesend again. Having discharged our 151 tons three days later we headed up river to Silvertown, where we loaded 162 tons of scrap iron for Hull. Although we reached Hull on February 18th we had to wait for a berth, and so it as not until the 25th that our last pieces of cargo left the Hold. Then it was off to Goole to load and back to Gravesend, where we arrived on March 5th.

Two Humber keels have finished loading under the Keadby coal chute on the Trent and are being hauled clear so that Everard's *Cambria* may drop back into the berth to load. Four tugs busy themselves in the river with a variety of craft including another Thames sailing barge. (Lincolnshire County Council, Gainsborough Library)

The next cargo for Hull was loaded at Millwall, Silvertown and Erith, followed by the return cargo of coal to Gravesend loaded at Keadby, and which was in the Thames by the 23rd. Leaving the coal berth we headed up river to Armstrong's Wharf at Millwall where we loaded a small cargo. There being no other parcels in the offing we had to sail with only 90 drums of black grease and 17 drums of oil, which earned the barge just £23.14s.0d. for the run north. We arrived at Hull on April 5th, discharged and returned to Gravesend coal laden by the 13th.

Although it was only early spring we had by then already carried twelve cargoes that year. And so the routine shuttle service went on, loading and unloading, keeping the holds clean and our vessel in good working order. April 22nd saw us arriving at Hull from Millwall again. After discharging, we carried oil cake from Hull to Wells. Returning to Barton on Humber we loaded 163 tons of tiles for Great Yarmouth. The tiles were discharged by May 10th and we then went light up to the Thames to load a further cargo for Hull at Silvertown and Millwall. Hull was reached on the 18th and ten days later we were once again at Gravesend with coal from Goole.

The next job was to load iron from McCall's Wharf up Deptford Creek. There we loaded 150 tons between the 3rd and 4th June and reached Hull by the 7th, returning to Gravesend with coal from Keadby.

As we were only able to go alongside the coal wharf on high water spring tides, I had to plan our passages from the north in order to arrive off Gravesend just at the right time, by so doing avoiding the need to anchor off. Our passage usually took thirty to forty hours, depending upon tides and the weather.

Although the months were slipping by, the pattern of trading was still the same, for on our return it was up to Millwall for oil and black grease again, then on to John Knight's at Silvertown for another few tons of black grease, all of which we carried to Hull, arriving there on June 22nd. By the 24th our cargo of 64 tons was ashore and fresh orders sent us on our way again to Burton Stather on the Trent to load more tiles for Norwich. It was to be an assorted cargo, with pantiles, ridges, half ridges, plain tiles and so on, in all about 85,000. The whole lot came to some 170 tons and earned the barge £55 gross freight against which expenses came to £11.9s.0d. By July 3rd these had been safely delivered at Norwich. We came away down the River Yare light, bound for Goole for our usual coal for Gravesend. It was July 15th by the time we arrived off Gravesend and were discharged by the 17th.

It was again the time for my wife and children to ship aboard for the school holidays. The boys loved their time spent on *Scone* and their mother was happy for us all to be together again. Usually they would join *Scone* when we were in the Thames, but sometimes they would travel to other ports to find us.

The Thames and Medway Canal in the early years of the century, with the little stumpy barge Pimlico of Rochester moored alongside. She was built in 1876. Dimensions of 73.5 feet overall with a beam of 14.5 feet would have allowed her to trade in the Regents Canal. (Kent County Council, Gravesend Library)

After I had spent a couple of nights at home they packed their things and we left before sun up for the walk to where the barge was anchored off. That was near the causeway at the Ship and Lobster public house, Denton, to the east of Gravesend. Our route took us down by the old Thames and Medway Canal, and I often told my boys about the days I used to sail model yachts there with my grandfather as a child. We went past Barton's timber sheds where the air was heavy with the scent of freshly sawn wood, mixed with the salty smells of the Thames which was shrouded in an early morning haze.

As we reached the far end of Barton's Wharf the made up ground gave way to the barge's graveyard, where rotting hulks lay in varying stages of decay. We walked on past the old *Mocking Bird* and another one with a kind of built up companionway. The old blue jerseyman was there as always puffing away at his pipe, no doubt with memories of better times past. At his side sat his only earthly companion, a large old black Labrador dog. As we left those hulks behind, victims of the declining fortunes of the sailing barge, I gave my usual signal, three long blasts on my pocket whistle, to call the mate ashore. Then the old reformatory Cornwall came into view, and beyond her several barges brought up awaiting a freight or a fair tide. We had arrived at the Ship and Lobster with the port's isolation hospital beyond. The quiet morning carried the sound of the *Scone's* boat splashing her way in as the mate's sculling made the clinker[1] hull clip into the still water.

As it was virtually low water we had the length of the causeway to walk, its lower end treacherously slimy with green weed. That always caused a few anxious moments until we were all safely aboard the boat. I had a word with the mate regarding the weather, and once on the *Scone* we got under way.

The boys had their favourite places aboard. In particular, Albert liked to sit down behind the bow rails, alongside the bitt head knees[2] of the anchor windlass. There he could hear the sound of the water being sliced through by the bows and also watch the sails filling in the wind. His other spot, if the weather was fine, was watching the wake as he sat on the lamp locker. There was a little shutter in the back of the wheelhouse so I could look from time to time to make sure he was not coming to any harm. Sometimes that perch of his was not so popular when the engine exhaust was swirling around. Their mother preferred them for'ard where she could keep her eyes on them, though they got used to the safe ways aboard at an early age. When the barge was on the wind deep laden the boys and their mother would sit on the lee side of the main hatch, dipping their feet in the water as the seas came aboard and ran the length of the deck before draining through the scuppers.

During the next three days we took on some 250 drums for Hull. Calling in at the wharves of Erith Oil, Armstrong's and John Knight's, we finally sailed from the London River to reach our destination and discharge there by the 24th July. That was followed by a run to Great Yarmouth with 40 tons of tiles from Barton on Humber. Then it was to London empty to load from the SS Clement for Strood. With 2,405 bags of cotton seed ashore by the 12th August, we were again on our way back up the Thames. After completing our topping up at the Erith Oil Mill we sailed northward to reach Hull on the 19th August, and a few days later we were deep laden with oil cake for King's Lynn. We returned to Burton Stather where we loaded 99,000 mixed tiles for Norwich. Part of that cargo was discharged at Great Yarmouth on our way through. By the 9th September we were back at Hull ready to load more oil cake for King's Lynn. We discharged our 167 tons on the 11th September and returned light to Goole, where 161 tons of coal was put aboard for Gravesend, where we arrived on September 22nd.

Deptford Creek was our first pick up after leaving the coal berth. There we took on 141 tons of iron, after which we added another 13 tons of general cargo from Millwall before making our run back down to Hull, arriving on the 4th October. That was followed by a trip from Burton Stather to Norwich with some 85,000 mixed tiles. Our discharging was completed by the 17th.

With that consignment our charter party stipulated that discharge should be carried out at a minimum rate of 40 tons per day, so a full freight could take over four days to unload.

[1] *Clinker is a stepped or lapped timber construction where the lower edge of each plank is outside the upper edge of the next, and so on.*

[2] *Massive timber knees fitted to the fore side of each of the bitts above deck to strengthen the windlass and spread the enormous loads to which the structure is subjected.*

119

St.Andrew's Waterside Mission church which bounds the east side of Bawley Bay, Gravesend. (Kent County Council, Gravesend Library)

This Bill of Lading for *Scone's* 157 tons 15 cwts coal cargo delivered on December 9th 1935 describes her as a motor barge. (Tony Farnham collection)

We went from Norwich to Boston where we loaded coal for Gravesend, leaving on the 23rd and arriving three days later, with empty holds by the 29th. Then it was down the coast loaded for Hull and still more coal to Gravesend, two weeks spanning the round trip, ensuring our return at the right time for the big tides at that most frequent of destinations. The coal wharf was sited just down river from the Town Pier. The wharf boundary to the east was at the sea wall which formed Bawley Bay, opposite the quaint little St.Andrew's Church. To the local residents of the town it became known as Jim O'Leary's, named after the proprietor who delivered to the houses of Gravesend. For myself, Gravesend waterfront was where my sailing barge career had begun all those years before.

Loading next took place for Hull at Millwall, Bermondsey and Strood, arriving at Hull with our 137 tons on the last day of the month. A few days later we were again heading south from Goole to Gravesend, where we arrived on December 9th. Our repetitive pattern of trade had us sailing up river to load at Silvertown and Millwall for Hull. We arrived back there on the 17th December. The cargo of just over 100 tons was out by the 18th. Our next orders sent us to Keadby on the 19th for more Gravesend bound coal. We reached the Thames on Christmas Eve but were not discharged until the 31st, which gave us a pleasant break. Once again it had been a good year for us, for we had successfully carried thirty-nine cargoes in spite of the bad trading conditions. Many of the other craft had not been so lucky. A lot of barges were laid up and their crews ashore without work. Many would never trade again, abandoned or broken up, too small or too old for profitable use.

DINHAM, FAWCUS & Cº LTD
8ᴬBILLITER SQUARE,
LONDON, E.C.3.

Weight—Weight shipped unknown.

CARGO.

Nº 1 Hold		tons
„ 2 do		„
„ 3 do		„
„ 4 do		„
Total	157 15/20	tons.

COALS ON BOARD FOR SHIPS USE, INDEPENDENT OF CARGO.

	tons in
	tons in
	tons in Bunkers.
	tons in Bunkers on arrival
Total	Tons

Shipped at GOOLE in good order and condition by MOTOR BARGE *Dinham, Fawcus & Cº Lᵗᵈ* in and upon the good ~~Steamship~~ called the "SCONE" whereof BAGSHAW is Master for this present Voyage and bound for GRAVESEND.

with liberty to sail without Pilots, to call at any ports in any order for bunkering or other purposes or to make trial trips after notice or adjust compasses all as part of the contract voyage.

a ~~cargo~~ parcel of One hundred and Fifty Seven ———— tons Fifteen ———— cwts

of NEWMARKET SILKSTONE Colliery, Railway, Dock Works weight, weight shipped unknown

which is to be delivered in the like good order and condition at the said Port of GRAVESEND unto J. O'LEARY ESQ. 31 PARROCK STREET, GRAVESEND.

or his Assigns, he or they paying Freight for the same as per Charter Party dated 193 all the terms conditions and exceptions contained in which Charter-Party are herewith incorporated.

General Average payable according to York-Antwerp Rules, 1924.

All the terms, provisions and conditions of the Carriage of Goods by Sea Act, 1924, and the Schedule thereto, are to apply to the contract contained in this Bill of Lading, and the Owners and the Charterers are to be entitled to the benefit of all privileges, rights and immunities contained in such Act, and the Schedule thereto, as if the same were herein specifically set out, the Unit under Article IV. (5) being the ton. If, or to the extent that, any term of this Bill of Lading is repugnant to or inconsistent with anything in such Act or Schedule, it shall be void.

In Witness whereof the Master or Agent of the said Vessel hath signed one Bills of Lading, all of this tenor and date, drawn as a set consecutively numbered, any one of which being accomplished the others shall be void.

December 4th. 193

CHAPTER 18

Distant Bells

"We left Strood to go down river to a large steam yacht laying at Upnor; she was going to Grays to be broken up."

1936 began with us laying alongside the SS Tulsa loading oyster shell grit for Grimsby. We finished stowing that new cargo late on 3rd January. We had found the work as a result of a chance meeting with a gentleman in the brokers office at Hull on one of our earlier visits. I had been asked about the possibility of bringing up a parcel of grit from London. I advised that our established clients had priority but, if the gentleman would like to accept my word, I would see that he was not let down. We were now harvesting the results of the seeds sown at that meeting.

Grimsby was reached on the 8th January and the discharging was completed the following day. We left light for Hull to load 167 tons of oil cake for King's Lynn where our cargo was discharged by the 15th, leaving us empty to return up the Humber to Goole for our coal for Gravesend. By January 23rd we were back on the London River and unloaded.

Over 3,000 bags of linseed were loaded into *Scone* from the SS Caduceus.
(National Maritime Museum, Greenwich, London)

Next we loaded three parcels for Hull and again returned with coal. The same recipe once more took us up to Leap Years Day, 29th February. Once emptied we went up to the Millwall Dock to load linseed from the SS Caduceus for Ipswich. Our 167 tons of cargo was made up of 3,096 bags, 19 of which were unfortunately damaged during unloading All were ashore by the 13th March.

From Ipswich we went down to Keadby for coal which we delivered at Gravesend on the 20th March. Millwall and Silvertown filled our holds for the return to Hull, arriving there on March 30th. Then we loaded caster meal for Whitstable and Strood. We sailed to the Medway first to discharge part of the cargo on 7th April and sailed on to Whitstable the next day to deliver the remainder.

While we were at Whitstable the wind was coming from the east and I knew that if we could get down to the Barrows we could make a good passage to Goole, where our next cargo of coal was waiting. Just as we were ready to sail, a messenger arrived with orders for me to proceed to Poole to load clay for the New Hythe Paper Mill. My immediate thoughts were for our regular customers and I felt strongly that we should maintain the Gravesend coal merchant's supplies, or that valuable business would be put at risk. I wasn't going to allow any intervention to jeopardise our hard earned regular runs of coal, just for one freight. I conveyed my thoughts to the owner's secretary quite forcibly over the telephone. As a result we were soon underway heading north to Goole to load coal. We arrived there on the 14th and four days later we were rounding up ready to go alongside the coal wharf at Gravesend. Our next freight was scrap iron from Deptford Creek to Hull under the financial arrangements we had made some months before. By the 29th our 165 tons of cargo was ashore and we proceeded to Keadby to load coal, that time not for Gravesend but destined for Orford in Suffolk, where we had loaded shingle in the past. We were discharged by the 6th May.

We then had to make a smart return passage to Hull to catch a ship which was unloading linseed. Our orders were to take 150 tons up to Ipswich at 7/- per ton. As bad luck would have it, there was a strong north-west wind blowing and we had the devil's own job in making our way north. We signalled the Lloyds Station at Spurn Point about 5pm, giving notice of our position and ETA[1]. Our efforts to get there paid off for we were able to load the linseed during the days of the 11th and 12th May, just saving the ship. After our discharging at Ipswich on the 16th it was back light up to Goole to load our regular coal for Gravesend.

We discharged by the 28th May, and once again found ourselves gathering cargo at Deptford Creek, Silvertown and Strood. It was the usual mixed load, 96 tons of scrap iron at 5/6d per ton which earned us £26.16s.7d. and then 58 drums of oil weighing 25 tons at 10/- per ton which brought in £12.10s.0d., a further consignment of 45 drums of oil at the same rate which grossed us £10.15s.0d. and finally 77 bundles of bags weighing nearly 3 tons at 14/- per ton. In all, the whole cargo earned the barge £64.19s.4d. against which expenses came to £8.3s.1d. It was all safely conveyed to Hull where we arrived on the 6th June.

In order to oblige a customer, the next voyage was quite a change of routine, for we carried cattle food from Hull to Shoreham and Strood. For a barge such as ours we again achieved a most satisfactory dispatch. Leaving Hull on the 19th June with the weather typical for that time of year, we arrived at Shoreham on the 22nd, a really smart passage, and discharged the following day. With still some of our cargo down below we made the return run fetching up in the Medway on the 25th to unload at Strood.

The next voyage was also quite different, not only on account of the passage but also with regard to the cargo. We left Strood to go down river to a large steam yacht, the Alacrity[2], laying at Upnor. She was going to Grays to be broken up. Her owner was transferring her fittings and furniture to another vessel down at Southampton. After loading up everything required, for a lump sum of £55 gross before expenses, we covered up and battened down with the wedges well driven home on the coamings. We sailed out of the Medway on June 27th to arrive at Southampton on the 29th.

We finished discharging the yacht furnishings on July 1st. Unfortunately we had to make the long return passage from Southampton to Boston empty. We got away before nightfall and brought up for the night in Ryde Roads. The following day we set out for Boston, arriving there on the 4th July after a non-stop run of forty-two hours. It was another very good passage, which made me feel I was running my ship as efficiently as a vessel could be run.

By the time we were ready to leave Ryde Roads on the morning of the 2nd, the wind was blowing strongly from the south-west bringing drizzling rain. I decided to shape a course to take us outside the Owers light ship. That meant holding to windward under the lee of the Isle of Wight. Once we had a good offing we bore away to run up Channel passing the Owers. We certainly did not require the use of our engine.

We shaped our course outside the Royal Sovereign lightship and up inside the Downs. Thereafter the south-westerly carried us rapidly outside the Kentish Knock and the Shipwash and on towards Southwold to pass down through the Stanford Channel and into Great Yarmouth Roads, on through the Cockle and down to Cromer.

As it was just coming daylight and the tides were in our favour, with the wind then backed fair to the south'ard, I shaped a course down through the bays. That took us close to Hunstanton pier from where we bore away over the top of the sands to Boston. As we were passing down that part of the coast we heard the bells ringing out from the churches along our route, a grand accompaniment as we glided along under sail.

Hunstanton Pier as it was in July 1936 when the *Scone* sailed close by before bearing away 'over the top of the sands' to Boston at the end of her non-stop run from Southampton. Less than three years later the pier theatre was destroyed by fire, forcing trapped holidaymakers to leap into the sea to escape the flames. The pier structure survived almost a further forty years until it was smashed to pieces by northerly storm force winds and seas in January 1978.
(Norfolk County Council, Hunstanton Library collection)

From Boston we returned to Gravesend where we arrived on the 9th July and were soon in the old routine with the family aboard for the summer holidays. Calling at Erith, Millwall, Silvertown then Strood, we sailed on the 16th July with a collection of general cargo totalling some 112 tons. We were back at Hull on the 19th.

All that voyaging between Hull and the Thames was costing us about £1 per week for fuel. The new engine ran on diesel oil and consumed about two and a half gallons per hour. Our fuel tank when full held 240 gallons, which gave us a nice reserve, even when I was compelled to use the engine all the way. At other times we could go from London to Hull and return to the Thames from Goole using as little as 50 gallons, being helped along by the wind and working the tides.

Scone at Burton Stather to load tiles. 'Harry's' sons, Cyril and Albert can be seen on the port side of the open hatch.
(Albert Bagshaw collection)

On July 22nd we were at Burton Stather on the Trent where we loaded tiles for Norwich. We took on a total of 80,000 plain tiles and 2,000 half ridges. After discharging we returned to Hull to pick up a cargo of linseed from the SS City of Roubaix for King's Lynn. The linseed was discharged by August 12th, following which we went back to Hull to load oil cake, part for the Wash port of Sutton Bridge and the remainder on up the River Nene to Wisbech. Thereafter it was back to Goole to pick up the coal which awaited us for Gravesend.

Planks have been laid to protect the deck and to make easier the movement of the barrow loads of tiles to the hatchways. The main runners and topmast running backstays have been taken for'ard to give unimpeded access to the holds. Inside the hold a heap of straw can be seen, which would be used to protect the tiles from breakages.
(Albert Bagshaw collection)

Kathleen always felt queasy when the *Scone* had an awkward motion in a seaway. It was perhaps not surprising for her background could not have been further from seafaring. She had come from gentleman's service as a parlourmaid before we were married. Although never seasick, a lumpy sea brought on an aching head. On that trip she came up from the cabin below to get some air and saw only turbulent seas and torrential rain all about us.

"How much further have we to go?" she asked. I told her that we would be coming to anchor shortly in the shelter of a sand bank to wait for a change in the tide. With that she returned below. A little later we dropped our anchor to wait out the adverse tide. With oilskins dripping I clambered down the companionway to find Kathleen looking a bit the worse for wear. "We're there." I said. "We can't be." she replied as *Scone* rolled and snatched at her anchor chain. Meeting her startled expression I tried to reassure her that the sea would soon be smoother as the tide fell, for we were were laying in the lee of the sand bank. Unbelieving she set off up the companionway. "I'm going to see that bank for myself." but of course she couldn't for it was below the surface of the sea. The uncomfortable motion of the barge soon diminished as the tide ebbed and she was herself again in no time.

After that freight had been emptied on August 21st, we came off the berth and headed down river on our way round to Frindsbury, up the Medway. We were due for a break while the barge went onto the shipyard for overhaul. As for myself, I took the opportunity of having a rare week's holiday away with the family, whilst the work on the barge went on at the yard.

The week passed quickly and it was back to work again on September 8th, taking the barge up to Millwall for loading. We loaded a fair sized cargo of acid oil and black grease which we carried north to Hull. Soft soap for Chatham Dockyard on the Medway was our return freight. Coming away from the Dockyard we headed down river and made our way up the Thames for our next cargo from Armstrong's at Millwall. We could only muster just over 41 tons for that trip to Hull but had a full load of coal for our return. It was September 30th by the time the last pieces of that cargo were swept and shovelled into the baskets and swung ashore and our holds were ready to receive the next cargo.

We loaded at Deptford between October 1st and 3rd almost 99 tons of scrap iron. Before leaving for Hull we topped up with a further 50 tons after calling at Millwall and Erith between the 3rd and 6th, arriving at our destination on the 9th. From Hull we carried 2,000 bags of barley meal to Ipswich, reaching there on the 15th October, and discharging during the following couple of days. Although we had orders for Boston our departure was delayed by a strong wind blowing from the north-west. However, we arrived at Boston by the 22nd and were loaded and on our way again by the evening of the 23rd. Our passage to Gravesend took until the 29th and we had emptied our holds by the end of the month.

We then went to the King George V Dock to load 196 tons of nitrate of soda from the SS Lautaro for Ipswich. The cargo came aboard over the days of the 3rd to 5th November and we reached our destination during the afternoon of the 6th. As usual, we were ignored until we had used up all the allowable lay days. It seemed to me that Ipswich freights were always urgent until we got there. By the

17th November we had completed loading at Millwall and Silvertown and were again ready for sea. The cargo was the usual mixture of merchandise used in the oil cake trade, chiefly oil in drums, plus bags of cattle food and various other stores. The freight came to £67 against which expenses came to £7.11s.6d.

We were at Hull discharging on November 23rd, then it was up to Goole and there loading 158 tons of coal for Gravesend. The first two days of December saw us empty again and the following days saw us making the usual round of calls at Millwall, Silvertown, Erith and Strood. With no chance of a single consignment to Hull which would fill the barge in those hard times, it was off hatches, on hatches, off hatches, on hatches, filling up the space below with small parcels to gather a cotchel freight for our run north. From Hull we repeated the routine of a fortnight earlier by going up to Goole and loading coal to keep the winter fires of Gravesend burning. It was December the 19th when we arrived back and discharging was completed the day before Christmas Eve.

That was the last trip for *Scone* during 1936. The twelve months had been well filled with useful work, for we had safely and successfully carried thirty-seven cargoes, as well as spending a couple of weeks on the shipyard.

This 'Pink Book' was issued on 21st December 1936 and on these pages are detailed some of the river freight rates for cereals and seeds, demurrage rates and definitions of 'ex ship' and 'ex mill', as well as minimum freight payments. For passages outside the Thames and Medway the 'Blue Book' provided similar information relating to coastal destinations. Both were published jointly by the Sailing Barge Owners Committee and the Transport & General Workers' Union.
(Albert Bagshaw collection)

CORN, &c.—*continued.*	Ex Ship. Per Ton. £ s. d. (20 cwt. to ton).			Wharf to Wharf. Per Ton. £ s. d. (20 cwt. to ton).		
Oats and Torrified Wheat	0	4	9	0	4	4½
Torrified Barley	0	6	8	0	6	1½
Quaker Oats, in cases ...	0	6	11½	0	6	5
Dried Grains	0	8	6½	0	7	10½
Decorticated Meal ...	0	4	3	0	3	11
Middlings	0	5	3	0	4	10
Bran and similar goods ...	0	6	2	0	5	8
Rice Meal	0	4	9	0	4	4½

Minimum freights : £11. 8s. for freights ex ship.
£10. 10s. for freights wharf to wharf.

After six clear weather working days for loading and discharging demurrage to be paid for at the rate of £1. 18s. per day for freights ex ship.
£1. 15s. per day for freights wharf to wharf.

No shifts to be paid for collecting goods in London.

When a barge is ordered to load *ex ship* the "ship" rate shall apply whether loading is effected direct overside or via silo. On the other hand, where the order is *ex mill* or *ex silo* the "wharf" rate shall apply, notwithstanding that a ship may be working alongside.

London to Faversham—

	Per Ton (20 cwt. to ton). £ s. d.		
Corn, &c., Wheat, Maize, Barley, Peas, Beans, Grain, Flour, Rice, all kinds of Seeds, loose or in bags	0	4	0
Oats and Torrified Wheat	0	5	0
Torrified Barley	0	7	0
Quaker Oats, in cases	0	7	4
Dried Grains	0	9	0
Decorticated Meal	0	4	6

26

CORN, &c.—*continued.* London to Faversham—	Per Ton (20 cwt. to ton).		
Middlings	0	5	6
Bran and similar goods	0	6	6
Rice Meal	0	5	0

Minimum Freight £15.

After six clear weather working days for loading and discharging demurrage to be paid for at the rate of £2 per day.

No shifts to be paid for collecting goods in London.

OIL SEEDS, &c.

	Ex Ship. Per Ton. £ s. d.			Ex Wharf. Per Ton. £ s. d.		
Oil Seeds, &c., to Rochester below Bridge.						
London to Rochester—						
Linseed, Egyptian Cottonseed, Rapeseed, Nigerseed, Soya Beans, Wheat and Seed Screenings	0	3	9½	0	3	6
Bombay Cottonseed	0	4	3	0	3	11
Ground Nut Kernel	0	4	6	0	4	0
West and East African Cottonseed, Smyrna, Mersene, Rice Bran, Maize Meal	0	4	9	0	4	4½
Soudanese Cottonseed, Brazilian Cottonseed, Gluten Feed	0	5	1	0	4	8
Ground Nut Meal	0	5	5	0	5	0
Uganda Cottonseed	0	5	3	0	4	10
Locust Beans, Malt Culms ...	0	6	2	0	5	8
Ground Nuts (in shells) ...	0	6	4	0	5	10

27

CHAPTER 19

**Trimmed
for Sailing?**

*"I had a heated argument with the dock manager over the payment for
our trimming...."*

Our orders for the opening of the New Year of 1937 were to carry scrap iron and general cargo from the Thames down to Hull. The scrap iron would be carried under our old and well tried agreement. It was loaded at Donald McCall's Wharf at the top of Deptford Creek. Like the coal wharf at Gravesend, it was only accessible on the spring tides. To get there we had to pass through a couple of bridges. The first of these presented no particular problem, but at the high railway bridge we had to top up the crosstrees and slacken up our standing and running rigging, unship the davits and lash the vangs in the middle of the barge, for the bridge opening was a mere eight feet wide. We loaded our 70 tons on the 7th and 8th of January, then we loaded a further 50 tons of general cargo at Armstrong's Wharf, Millwall.

Hull was reached on the 17th January and there followed another southbound passage to Gravesend with coal. Discharged by February 5th, we made our way round to Strood to pick up our next cargo for Hull. That was a straight forward 159 drums of black grease at 14/- per ton. The run northward was completed by the 15th and with our cargo safely delivered we sailed for Goole to load coal again for Gravesend.

While we were discharging on the 23rd January I received our next orders. They were not very encouraging, for the quantities of merchandise offered us for Hull were now dropping off even more. Although we made calls at Strood, Millwall, Erith and Purfleet we only loaded 100 tons. Hull saw us on the 8th March, the freight grossing us £62.9s.0d. against which our expenses came to £7.0s.9d. While at Hull we loaded 170 tons of oil cake for King's Lynn, where we eventually arrived on March 13th. Returning to Boston we loaded coal for Gravesend, unloading on March 20th. From Gravesend we went down to Strood Oil Mills to pick up 48 tons of compounds and 12 tons of oil for Ipswich. As always we did not get a quick turn around at Ipswich, for although we arrived on the 26th March we were not able to discharge until the 30th. Leaving the River Orwell our heading was then north to Goole empty to pick up 156 tons of coal for Gravesend.

Arriving back at Gravesend on April 7th, our discharging was completed two days later. Making the most of the tides, Deptford Creek was our next loading place. Under the hatches went just over 100 tons of scrap iron from McCall's for Hull, carried for a freight rate which had just been increased by sixpence per ton. Topping that up to a full cargo at Millwall, we set out on our northward passage, arriving at Hull on April 20th.

Then orders came which brought a slight change in our routine, for instead of us picking up the usual coal cargo, we went higher up the river to Stoneferry where we loaded 106 tons of soya meal for Strood, discharging there by the 26th. We immediately began loading from the Strood Oil Mills for Hull. Before sailing northward however, we went to top up at the

Younghusband, Barnes & Co.'s wharf at Bermondsey. Leaving the upper reaches of the Thames on the 28th April, we arrived at Hull to start discharging on the 3rd May. That time we took the familiar course to Goole, where coal awaited us for Gravesend, *Scone* unloading there between the 13th and 14th May. We again made a round of calls picking up cargo before starting out for Hull. These visits included Strood on the Medway, then Deptford, Millwall and Silvertown on the Thames.

For all our moving around there were only small quantities offered at each of these places. The whole cargo only came to 80 tons, the gross freight worked out at £42.7s.11d. which was not much return for our labour and time. It was the first day of June before we were finally ready for sea. However our departure was held up due to the owners wanting my services for the Thames Sailing Barge Match. That year the race was run in celebration of the coronation of King George VI, and our firm had appropriately entered just two craft, *King* in the Coasting class and *Queen* in the Champion Bowsprits. Neither managed a victory, but both were runners up to the top racers. Everard's *Veronica* beat *King* by almost half an hour. *Queen* worked her way into the lead on the way back from the Mouse, but was eventually overhauled by Everard's *Sara* which went on to win by just under ten minutes. Nevertheless *Queen* did have the satisfaction of taking the prize for the fastest start. Despite my views on racing I could not help but get caught up in the atmosphere of that special occasion and my commemorative medal became a treasured possession.

'Harry's' cherished medal commemorating his participation in the Coronation Thames Sailing Barge Match of 1937. Note that the medal bears the inscription 'RACE' and not 'MATCH'. (Albert Bagshaw collection)

Opposite:
Scone lies at King's Staithe in the Purfleet at King's Lynn, alongside Vynne & Everett's ancient warehouse. The 'turret' on the corner of the warehouse was where the crane operator sat. It was also a good vantage point for spotting approaching craft, with a commanding view downstream. (Albert Bagshaw collection)

Scone did not get to Hull until the 9th. It was strange that when the freights were poor, the passages always seemed to take longer. However, we discharged the following day and were sent straight away south to the Thames empty to pick up an urgent cargo of 100 tons of bone meal at Silvertown. We headed north again, a little more than half laden, and by the 21st June were back at Hull. That earned us a £60 freight for our trouble.

From Hull we loaded linseed alongside ship, which we carried to King's Lynn. Leaving Lynn light we went to Boston to collect coal for Gravesend, which was discharged over the 2nd and 3rd July. Our next orders had us on our way to Strood again, before making the usual round of calls in the Thames, picking up small parcels of cargo for Hull. After discharging there on the 19th July, we loaded 165 tons of maize which we carried to Great Yarmouth.

Albert swings in the bosun's chair whilst *Scone* visits Kings Lynn. (Albert Bagshaw collection)

Sailing north to Barton on Humber empty, we loaded 165 tons of tiles for Rotherhithe, London. Our passage south to the Thames had us arriving at our destination on July 29th. With the bank holiday interrupting our unloading, we were not ready to receive the next cargo until August 3rd. Disappointingly, the amount offering for the northward run was very small. After calling at Millwall for 24 drums, and Erith for a further 28 drums we went to Hull with a bare 22 tons, just over £11 for the whole cargo before expenses. We had one consolation; our full cargo of coal would be waiting for us when we reached the Humber, ready for delivery to Gravesend. That was followed by us going up to the South West India Dock where we loaded 3,220 bags of oyster grit for Hull. Filling our holds during the days of the 19th and 20th August from the SS Llanwern, we made our way northward with our cargo of 161 tons which was discharged on August 24th. We were immediately refilled with 1,688 bags of linseed loaded from the SS Calcutta. We made the passage to King's Lynn by the 26th, where the hold was empty again by the 27th.

Our stop at Lynn was not without incident. The family had joined the barge some weeks previously and that day Albert swung in the bosun's chair from the burton tackle on the sprit whilst his brother decided to do a spot of fishing. After tackling up Cyril suddenly cried out; he had got his fishing hook through the skin of his finger. The nearest help was the local police station where luckily the police surgeon was in attendance. He

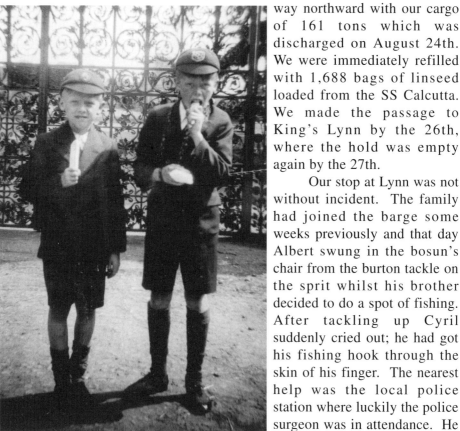

Cyril, sporting his newly acquired finger bandage, with Albert outside the gates of Sandringham later on the day of the accident. (Albert Bagshaw collection)

removed the offending hook, and with a suitably impressive plaster and bandage we were soon on our way.

After running back light to Hull, we loaded fish meal for London. Sailing south we arrived back in the Thames on September 4th and had discharged our cargo of 160 tons at the Dundee Wharf by the 7th. The freight rate was 6/- per ton giving us £48 gross.

We made calls at Millwall, Silvertown, Rotherhithe and Queenborough, the latter a new place for offering a cargo to us. In spite of all these calls and the different bundles of cargo we were carrying for different merchants at the other end, we were loaded and ready for sea by September 11th. At Silvertown we had loaded 24 drums of black grease for one merchant and a further 32 drums for another while we were at Millwall. Then there was a consignment of 89 barrels of acid oil loaded aboard at Rotherhithe, before completing our cargo at Queenborough with 50 tons of meal.

Quite apart from the running of the barge, the care of all that assorted merchandise and the responsibility for ensuring that it was all delivered to the correct consignee's was quite a job, yet the whole freight amounted to only £41.4s.0d. How valuable our regular southbound coal work had become; without it I would have had great difficulty in gaining a living wage to support my family.

Arriving at Hull on September 16th, we managed to discharge all our parcels in one day, before going to Goole to collect our coal. We were unable to find any cargo offering for the next northward run, so after discharging on the 22nd we went back down the east coast light to Boston. There we again loaded coal for my home town. Our orders called for 133 tons of cargo to be 'trimmed' away below. Drawing on my previous experiences, I had my suspicions about the extent of that operation. The way of loading certainly looked as if the coal was just dumped down into the holds without any trimming being done. Trimming was a job that we had to pay for, and of all the coal ports we visited, Boston was one of the worst, for it had never been carried out correctly. The fact that the task was improperly done deprived us of a good deal of our hard earned freight money and could have put our lives at risk if the cargo had shifted, more especially so being a sailing vessel. These facts were quite immaterial to the labourers who undertook the work. Little did they realise that we would be doing the unloading at the other end and could see for ourselves their appalling workmanship.

I had a heated argument with the dock manager over payment for our trimming, and complained bitterly about the men who were not doing what they were paid to do. For all my seamanlike concern for my ship, we didn't seem to be getting anywhere on the subject. There seemed only one way to illustrate my point to him and his men. Seizing a round mouthed shovel I went aboard my barge, which according to them was loaded, trimmed and ready to sail to sea. In full view of the dock manager I soon dug down into the coal at the after end of the main hatch, over towards the port side which was the off, or furthest away, as the barge laid alongside the quay. The length of the hold was 66 feet. I was no dwarf at 14 stone but, once deep enough, I slid under the coaming into the cavity space under the deck. Without much trouble I was able to crawl forward toward the bow and out of the fore hatch. That somewhat drastic demonstration proved that our cargo could have shifted when the barge was pressed down sailing and risked us losing the ship and our own lives. As soon as I emerged, a little dirtier for my trouble, there was no need to say more. The immediate result of my demonstration was that the trimmers resumed with added instructions from their manager that it was to meet the skipper's satisfaction before they left the job. After that episode we noted a marked improvement, but probably only so far as our craft was concerned.

Following our arrival at Gravesend on October 2nd, it was the 5th of the month by the time we were ready again to receive cargo. That took the form of another round of calls at Strood, London and Queenborough. We were fortunate in picking up quite a sizeable cargo that time which earned the nice sum of £65 gross.

It had its drawbacks, however, for it was a very mixed cargo and required a great deal of careful judgement in order to keep everything in the right place so that it could be got at in the right order, yet play its part in establishing the correct and safe trim of the barge. There were 44 drums of oil weighing nearly 19 tons and 76 bags of cotton press bagging, weighing 3 tons. Another 9 tons was made up by 22 drums of grease, while a further consignment of 46 bundles of cotton press bagging weighed nearly 2 tons. Then came another two consignments of grease in drums, one lot numbering 54 drums and weighing just over 23 tons and the other lot numbering 58 drums weighing almost 25 tons. Finally there was 50 tons of meal which was picked up at Queenborough before we battened down and set out for Hull.

After discharging that very mixed cargo we carried 160 tons of oil cake from Hull to Wells. That was followed by a return run to Goole to load coal

Topsail barges lay at Kingston-on-Thames alongside coal filled lighters. An ex sailing barge hull, stripped of spars, sails and rigging, is in use as a lighter to the left of the photograph. (Peter Ferguson collection)

bound for Kingston, which took us right up the Thames under all the London bridges. It made a refreshing change for us, although it gave us a good deal more work and worked out less profitable than delivering to Gravesend. Going up under the bridges naturally necessitated the lowering and upping of our gear, several times, as well as dropping the boat and unshipping the davits. At Teddington I had some anxious moments for our mast only just cleared the underside of the bridge.

On the 6th November, after making our way down the river, we swung alongside Armstrong's Wharf at Millwall to begin loading for Hull. Before we had enough cargo to put to sea it became necessary to make calls at John Knight's Wharf at Silvertown, then Rotherhithe and finally Strood. Eventually we had aboard what I considered was a worthwhile freight and so we set off down the river and out of the estuary past the low lying shores of Essex, Suffolk and Lincolnshire and on to the Humber. It was not until 17th November that we were again entering Hull. Discharging went a good deal quicker that time, however, and by the 19th we were at Goole loading 155 tons of coal for the return passage.

On November 22nd we arrived at Gravesend and had discharged a couple of days later. It was time to pay another visit to Deptford Creek, where 100 tons of scrap iron awaited us. Leaving the creek behind we made our way to Millwall for a small quantity of cargo before calling in at Erith for 40 tons of oil in drums. We left for Hull just before high water, a passage I knew like the back of my hand. It was not quite so simple as that however, for it was November and the weather was bad; so bad in fact that we were forced to put into Great Yarmouth for shelter.

Eventually Hull was reached on the 8th December and we had discharged by the 11th, earning a gross freight of £59 with expenses coming to £7.17s.6d. We spent the next three days loading at the Eagle Mill with 147 tons of oil cake for the relatively short passage to Wisbech, where we arrived on December 15th. Although we had completed unloading by the 18th and had earned £38.12s.0d., there wasn't much profit, for our expenses were high, coming to £10.16s.9d.

The year came to a close with us running light to Goole and returning after a good passage to Gravesend with coal. It was Christmas Day when we arrived in the Thames, coming to anchor off the Ship and Lobster at Denton. We were soon ashore, and it was a pleasure to be able to spend a couple of days with my family.

With the festivities over, the coal wharf was ready to receive us. We made our short trip up river and berthed to discharge, the final baskets swinging ashore on the last day of the old year. It had been another good year for us in spite of some slack times and the struggle we often had to gather a worthwhile freight for the run north.

We had also gained a dog. She was a little black mongrel bitch; a stray which had hung around the barge at one of our berths until the mate took pity on her and she was taken aboard. It was decided that the new arrival should be called Fluff and she soon got used to the ways of the sea, clambering up and down the steep fo'c'sle ladder, sure footed around the deck, and waiting by the companion way aft when it was time for her food.

CHAPTER 20

Lull Before
the Storm

"... we proceeded up the Breydon Water to the Haddiscoe Cut, which seemed composed of slimy mud rather than water."

New Years Day, our first day of trading in 1938, found us alongside a steamer in the South West India Dock loading 3,100 bags of oyster shell grit. The SS Lionstancl had arrived from the United States and that portion of her cargo was carried by us to Hull, arriving on January 7th.

From there we carried just over 149 tons of oil cake to King's Lynn and then, as often before, we went over to Boston to load coal for Gravesend. That was discharged by January 20th and by the end of the same day we had shifted berths to tie up alongside a wharf at West Thurrock, where 160 tons of cement was loaded for Great Yarmouth and up the Yare for Norwich. The rate for the cement was 6/- per ton which brought in £48 with expenses coming to £8.12s.1d. After landing the cement we made a light run up to Goole, where we arrived on February 5th, returning to the Thames to unload our coal at Gravesend on the 9th.

The next lift was for general cargo which was picked up at various places in the Port of London as well as a call at Strood in the Medway. Trade seemed to be improving. It was a big cargo for Hull, as big as we had handled for a long while. It brought in £76.15s.7d., but it must be remembered that although the freights were better the expenses were rising too.

The return passage was made from Boston to our berth at Gravesend, where our coal cargo was worked out with the use of our motor winch, using three hundredweight wicker baskets. Our usual cargo of just over 150 tons could be discharged in a day and a half.

The nitrate of soda cargo loaded from the 1919 built SS Loreto was vulnerable to damage by moisture. Barge and ship had to cover the cargo in showery weather every time there was a hint of rain; a frustrating process which inevitably delayed both vessels. (National Maritime Museum, Greenwich, London)

By the 9th March we had completed taking in from Millwall, Silvertown and Purfleet, gradually building up our cargo before finally battening down the holds and making ready to put to sea for the voyage to Hull. Our return coal was discharged by March 22nd.

On the following day we went into the Royal Albert Dock to load nitrate of soda for Great Yarmouth from the SS Loreto. Owing to prolonged periods of continuous rain, which could not be allowed to damp the cargo, we were not finished until the 30th. However, we entered Great Yarmouth during the evening of the following day and were discharged a couple of days later.

Running back up the Thames light, we loaded a consignment of cased oil at Millwall and Silvertown for Hull. As there was nothing else offering for the Humber ports, we went down to West Thurrock and took aboard 50 tons of cement for Great Yarmouth. By April 6th we were ready to make our passage down Northfleet Hope and out to sea.

We made our call at Great Yarmouth on the 9th and arrived at Hull on the 11th. From Hull we received orders to go up the river Trent to Burton Stather where we loaded tiles for Norwich again. It took three days to load the 168 tons of cargo, for which the rate was 6/6d per ton, making our gross earnings come to £54.12s.0d. with expenses taking away £12.2s.0d. Our next assignment was to carry 170 tons of shingle from Orford Haven to Grimsby. It was not a good freight rate at only 3/6d per ton, with expenses coming to £5. However, it took us in the direction we wanted to go, and we were soon fully laden with coal again for the run south.

Over the 18th and 19th May we laid alongside the SS Andres in the Royal Albert Dock taking on board 151 tons of oil cake in bags. That was carried to Great Yarmouth and instead of continuing northward we returned light to Queenborough, where we loaded 50 tons of meat meal for Hull. That call preceded our usual round of calls at Millwall and Erith, which in so doing, provided a full cargo for the Humber. Leaving the Thames on the last day of May, we arrived at Hull on the 4th June. Our discharging was drawn out that time, for we had three berths to visit, taking us until June 9th.

From Hull we went over to Barton on Humber where we loaded more tiles for Norwich. 81,000 pantiles and 1,000 'Lynn' ridges were stowed down below in the hold, with plenty of hay and straw for packing. That was a 165 ton cargo with a rate of 6/6d per ton. We arrived at Norwich on June 13th and were sweeping our empty holds by the 15th.

From Norwich we returned light to the Thames. In the West India Dock there were 3,000 bags of oyster grit awaiting us, which we took to Hull, where we arrived on June 26th. Our discharging took place the following day. There followed another coal cargo from Boston to Gravesend, where Kathleen and the boys joined the barge for the school holidays. Once again we made a round of calls at Millwall, Silvertown and Queenborough, much to the excitement of the Cyril and Albert who were fascinated by the bustle of the upper reaches. The tugs, lighters, steamers and of course the sailing barges, seemed to fill the river, yet to their amazement, or perhaps even disappointment, there were very few collisions.

The barge's total earnings for the trip came to £65 with expenses of £8.15s.4d. With our passage made without incident our discharging was completed at Hull by July 13th. Burton Stather was to be our next port of call to load tiles for Great Yarmouth. On that occasion I had accepted a rate of 5/9d per ton, on the understanding that our loading and discharge took no more than four days. My Director gave me a reprimand for reducing the rate, but I told him that we were limited to time and the saving of three days in discharging was well worth the 3d discount.

Our destination was reached with 171 tons aboard bringing in £49.5s.7d. less expenses of £7.16s.5d. As usual when at Great Yarmouth our discharge berth was up towards the swing bridge end of the harbour on the west quay. When we had finished for the day we all went ashore and made our way down to where the ferryman ran his rowing boat across the water to the east quay. As we crossed the river we could see the substantial fleet of herring drifters laying together head on to the quayside and behind them all their nets hung up to dry

The west quay below the swing bridge at Great Yarmouth with a barge, wherry and what may be a topsail schooner alongside. Another flush decked coasting barge lays outside the wherry, which appears to be using its counterbalanced mast as a derrick for lifting a bagged cargo from her hold.
(Tony Farnham collection)

and for mending. Our destination was the amusement park on the southern end of the promenade, quite some walk for the boys. However that was not a problem on the way there, but Albert my younger son spent much of the return journey carried on my shoulders. From Great Yarmouth we turned north to Goole light for our coal for Gravesend. The time saved earlier on enabled us to reach Gravesend by the 25th, thus saving our water on the berth and allowing completion of our discharging before the tides began to take off again.

Scone's departure from Gravesend was in the early hours of 28th July. Our orders were to reach Strood on the Medway by the next high water, as it was our turn to go on the shipyard for a refit. Arriving about half ebb, we anchored to await the next tide. That gave us time for a particularly thorough clean up, washing out the holds and sweeping off the tarpaulins. Once on the yard we put all our spare rope and other gear down the hold alongside the keelson. The barge was maintained in every way in first class condition, nothing was left to chance. I presented a list to the foreman shipwright advising what repairs needed to be done. The engine was overhauled by the yard men. The sails and rigging were left to me to dismantle with the help of the mate. Top of our replacement list was a new suit of sails. All the small wires were replaced, all running ropes were renewed, and so were all the shroud lanyards. Whilst the barge was on the yard I lived at home, taking one or other of the boys down every other day to Strood.

One day Albert had been hanging about waiting for our boat to float. When it did he sculled back and forth in the river like an old hand, despite the fact he was just eight years old. The sailmaker came to tell me that our new topsail was ready to be collected so I called to one of the men nearer the river's edge to tell Albert to fetch it from the sail loft. He needed no second bidding and sculled the boat to the landing stage at the loft and made himself known. They loaded the sail for'ard as was usual, forgetting that there was just a small boy aft to counterbalance the weight. But Albert made it back safely, thinking himself a true bargeman. The refit lasted just over three weeks, during which time I managed a rare weeks holiday away from the barge.

The company's bargeyard at Frindsbury was once Curel's before passing into the ownership of Gill and Sons, then The Rochester Barge Company, later The London & Rochester Barge Company, then The London & Rochester Trading Company, then Crescent Shipping and finally Hay's Marine Services as the firm's trading style and ownership evolved. The yard was sold in 1996 to new owners and faces an uncertain future in the closing years of the century. Here we see the yard in its heyday, with the paddle steamer Princess of Wales on her way down river. (Bob Ratcliffe collection)

Our first charter after leaving the shipyard, with *Scone* looking like a new pin, was to carry 752 quarters of wheat to Ipswich from the Surrey Commercial Dock in London. We got a big surprise on our arrival, being discharged on the very same day that we had arrived, a unique occurrence at Ipswich! As a result we were able to arrive at Goole on August 28th to load 155 tons of coal for Beccles. Beccles was no easy place to reach. After entering Great Yarmouth, we proceeded up the Breydon Water to the Haddiscoe Cut, which seemed composed of slimy mud, rather than water. However, we arrived at Beccles on August 30th and had been discharged by the first day of September. Afterwards we went light down to Boston where another 155 tons of coal was loaded for Gravesend.

Having discharged back in the Thames, we spent the period from the 15th to the 20th September making calls at Silvertown, Millwall and Queenborough loading general cargo for Hull. There was a large amount offering and we collected a total of £74.11s.6d. in freight, with expenses coming to £8.15s.5d. With our cargo safely ashore at Hull, we made the usual return passage from Goole with coal, which we discharged on the 28th September.

137

By that time the Country, and indeed the World, had just passed through the harrowing events surrounding the Munich Crisis. We carried on as usual, however, practically unaware of what was happening within the international field of politics.

A short run followed with 160 tons of cement from the Gillingham Cement Works. We had to lower our mast for that freight on our way up to Fulham, where our cargo was discharged by October 6th. Once again we commenced the usual round of visits sailing to Millwall, Silvertown and Erith, picking up cargo for Hull where we arrived on October 13th. From Hull we loaded 146 tons of soft soap for Chatham Dockyard, for which we received the lump sum of £45 with expenses only coming to £5.3s.1d. It can be seen that the miles covered by *Scone* were adding up to a staggering distance. Voyage after voyage was made in unending procession in spite of all the hazards of fog, gales and strong tides. Fortunately, I had a dead true compass and a reliable log to say nothing of the lead line, which we always had in use.

The discharge of the soft soap at Chatham was completed on October 18th and was followed by orders which sent us up to the Surrey Commercial Dock to load 647 quarters of wheat from the SS Canada for Felixstowe. The freight rate was just 10d per quarter so the voyage only grossed £26.19s.2d., less expenses of £3.11s.10d. That was followed a cargo of 815 quarters of maize for Great Yarmouth which we loaded out of a ship at Butterman's Bay on the River Orwell, just below Pin Mill. For that we received £26.3s.6d. less £3.15s.2d. expenses. From Great Yarmouth we ran to Goole where we loaded 153 tons of coal for Gravesend, arriving there on November the 6th.

At that particular time the improvement in cargoes for the run down had not continued and there was very little merchandise about. We had only secured ten tons which was loaded at Millwall. However, just as we were about to sail we picked up 677 quarters of wheat for Felixtowe from the SS Llanwern laying in the Millwall Dock. We made the call on our voyage north to Hull. It was November 28th when we arrived back in the Thames with 160 tons of coal from Goole.

All the while our engine had been behaving perfectly. I made a practice of having our fuel topped up at Purfleet when entering or leaving London, as time and tide permitted. Although our consumption varied, generally the round voyage, London to Hull and back to the Thames, would use 150 gallons. It depended on the conditions of tide and weather; sometimes more and sometimes less.

On 30th November we went up to Millwall to start loading for Hull and finished down at Queenborough on December 3rd with 100 tons of meal. On that voyage we had about 140 tons of cargo aboard which brought in the excellent sum of £67.1s.6d., less £8.11s.1d. for expenses. From Hull we carried oil cake to King's Lynn and arrived back at Hull on 14th December where we picked up a similar cargo waiting for shipment to Wells. After taking aboard 135 tons, we put to sea on the evening of December 15th.

Right from the start conditions were pretty bleak and wintry with strong easterly winds and frequent showers of snow as we proceeded down to Spurn Point, where we anchored for shelter. After laying there for some days I decided to go back to Hull and await a lull in the weather. On our return I contacted the shippers and they took the opportunity to load another 20 tons of cargo. The weather showed absolutely no signs of letting up so we put the *Scone* into the

William Wright Dock and tied up for the Christmas holiday. Christmas was spent at home and I did not get the train back to Hull until the 27th. The weather had become a great deal kinder, at least in strength, if not direction, so we put to sea.

The harbour at Wells was by no means easy to find and by no means easy to enter either. It was a blind place; by that I mean that there were no leading lights, just a spar buoy off the entrance as a guide, so that with strong tides and bad winter weather it demanded a deal of good judgement.

As we came out of the Humber the wind was still easterly which was always bad for Wells. But it had moderated a lot since our last attempt, though was joined by a little fog. I had intended going up the Wash to await a favourable chance to get into Wells. However, I changed my mind and set a course for the Dowsing Overfalls buoy, and then a course for Wells, making a good allowance for the strong set of the tide from the south-west. It grew dark about four in the afternoon and I had begun to be doubtful of the wisdom in closing Holkham Bay. When I considered that we had travelled far enough, we anchored for the night, taking great care with the lead to ascertain our position, for the water shoaled considerably there. We were laying in a very exposed anchorage so we had to keep anchor watch just in case the wind freshened. Fortunately it died away, its place being taken by fog. Later when the moon rose the fog cleared and in the silvery light we discovered that we were only a ship's length away from the buoy.

It was December the 29th when we arrived at Wells with what was our 38th cargo of the year, although it was not until the 2nd of January 1939 that our holds were completely emptied.

The drying harbour of Wells-next-the-Sea on the Norfolk coast serves the local agricultural trades. The *Alf Everard* is alongside with her bowsprit lowered, perhaps waiting to leave on the tide. Wells has in recent times received cargo by auxiliary sailing vessel, the Dutch klipper barge Albatross delivering soya meal for a number of years until September 1996.
(Peter Ferguson collection)

CHAPTER 21

Troubled Times

"We were forced to navigate slowly, ever on the look-out for mines; moreover we could only sail during the hours of daylight."

Three days of the New Year of 1939 had already passed by the time we left Wells. With the wind fresh from the north-east, almost straight onto the land, it was piling up a considerable swell. After a lot of thought I decided that we should be able to get out of the Harbour, and if successful in the venture, carry on down by Scald Head, then through the bays to Hunstanton and finally across to the Boston Channel. With the aid of our sails, which were set while still in smooth water, and our Kelvin engine, we came out nicely. A few short tacks were required and the worst was soon behind us. *Scone* did well considering the strong wind, the savage easterly tidal set and the fact that she was a light ship.

January saw us back in the pattern of scavenging for parcels of cargo in the Thames and Medway before heading down to the Humber, then back with coal to complete the cycle. During that first month of the New Year *Scone* had earned £154.9s.10d gross. Total expenses to be deducted came to £22.19s.9d. leaving a net income of £131.10s.1d. In the customary fashion, the crew's share of the money was £65 from which we had to pay the cost of the fuel used in our engine and motor winch. That amounted to £2.9s.4d. It is interesting to note that the fuel oil bill for the whole year of 1938 came to £40.13s.3d. and that included the petrol for starting the engine.

Scone's Half-Yearly return endorsed 3rd January 1939 in Boston.
(Albert Bagshaw collection)

Leaving Gravesend at the beginning of February, we proceeded up river for the West India Dock. There we loaded 100 tons of oyster grit from the SS Tulsa before carrying on up to Millwall to load more cargo. Finally we fetched down to Erith, where the last parcel came aboard to be stowed in the fore hold just under the mast case. As I looked on I heard a sharp crack from aloft. A man from the Erith Works who was assisting in the loading was just outside the coamings unaware of any danger. In a split second I grabbed the astonished individual and pulled him clear just as the topmast came crashing down. The crane jib had fouled our topmast stay, breaking off the topmast and bringing down with it a mass of tangled rigging. Our helper was clearly shocked by his narrow escape, but quickly recovered. As a result of the damage we had to go on the yard at Strood to have a new topmast fitted.

While our gear was down I took the opportunity of going over it pretty thoroughly and found that the peak end of the sprit was in a bad way. So we also had a new sprit and although it was only an inch smaller in diameter than the one it replaced, it appeared slight, something I did not like, bearing in mind the hard winter passages we undertook when other barges lay waiting fairer weather. Another thing which I was not happy about was our starboard leeboard, which I considered was no longer strong enough for the stresses it had to stand. I voiced my concerns, saying that we should be in trouble sooner or later, when we encountered the next real breeze.

The repairs at Strood only took a couple of days and by February 6th we were down at Hull again. Between the 8th and 9th of February we loaded 160 tons of tiles at Barton on Humber which we carried to Great Yarmouth, where we were emptied by February 15th. Returning to Goole we filled *Scone* with coal which we delivered to Gravesend on the 20th.

Then came a round of calls, firstly at Strood, then back up the Thames to Millwall and Erith, where I made sure that the crane jib would not foul the short topmast stay a second time. We finally completed at Queenborough and were off to Hull, reaching our berth on March 3rd.

It was quite a large freight and it brought us £72.18s.8d. with expenses totalling £10.9s.7d. Our southbound coal freight was from Boston and was followed up by 767 quarters of wheat from the Surrey Commercial Dock round to Ipswich where we arrived on the 11th, before going on light to Goole. Owing to extremely bad weather we did not reach Goole until the 24th and it was April 4th before we were back at Gravesend.

Millwall, Silvertown and Queenborough were first visited to make up the next run north. Whilst gathering many parcels for our trip we also found ourselves at Strood and took the opportunity to have the engine de-carbonised. It needed doing after each 1,000 hours of running time and the valves were re-ground in at the same time. Before going north we went up to Erith to pick up the last of our cargo. We arrived at Hull on April 21st.

We carried 160 tons of maize to Boston for £32.3s.9d. less expenses of £7.4s.7d. and then returned to Hull to load oil cake for King's Lynn. We arrived at Lynn with over 150 tons on the 29th and finished discharging by May 3rd. Then it was a light passage to load at Goole and we were back at Gravesend on May 7th. The rate for carrying coal was usually 6/2d per ton. With 155 tons on that trip we took in £47.17s.4d. and paid out £8.4s.8d. expenses.

Once again we made the usual round of loading places, calling at Silvertown, Purfleet and Queenborough, and by May 16th we were back at Hull. The return passage to London was made with 120 tons of flake meal which was for Bow Creek. All went well until we were coming up the Thames off Erith, when the propeller shaft broke. We were forced to complete the passage under sail. When fitted, the propeller shaft had been lined up when the barge had been light. The new shaft was installed and lined up when the barge was loaded, which meant that less stress came on it, and in fact it was never a problem again. After we had taken on cargo at Erith and bunkered at Purfleet, we sailed for the north.

Leaving Hull we carried a cargo of poultry food up to Ipswich discharging there on the 13th June. We sailed on to London empty and we were ordered to the West India Dock where 3,000 bags of oyster grit were loaded from the SS Tulsa again. It was June 20th by the time we had arrived at Hull and

discharged, and it was five days later when we were back at Gravesend with a cargo of coal from Boston, though it was the 28th before there was enough water for us to get alongside. As soon as our holds were empty we went up to the Millwall Dock where we filled *Scone* with 139 tons of cotton seed for Ipswich, returning light to the Thames on July 7th.

In Tilbury Dock we loaded 143 tons of linseed from the SS Maihar, again for Ipswich. The rate on that freight was only 5/10d per ton. Instead of returning south we carried on light to Goole from where we brought back the usual cargo of coal to Gravesend. All the time I had been getting more and more apprehensive about the condition of our Sprit. I had made known my worries to the owners, who in response spent quite some time trying to convince me that it was a strong enough spar. Nevertheless I remained unconvinced, and time would tell who was right.

Between the 22nd and 27th July we lay in the West India Dock loading more oyster grit for Hull, but before putting to sea we also made calls at Millwall and Purfleet. Once discharged on August 1st, we went up to just inside the Dutch River at Goole to load 170 tons of potash for Boston. The tidal current ran very strongly in the Dutch River, so strong in fact it shook the bridge that spans the river entrance. We were forced to tie up alongside a pier to await the bridge being swung open.

I deduced that to hang on until high water would be against us. To pass through early on would be best, giving the opportunity to put the barge's stem into the mud to stop her. With that in mind I clambered ashore and walked round to the bridge master's office. At first he refused to open the bridge until the slack at high water, but eventually I persuaded him that it would be quite all right to let us through. Back aboard, the moment the bridge opened we let go. With the full weight of tidal surge behind us we swept on through the forty foot wide bridge hole at nearly ten knots, and rounded easily to our berth. That enabled us to start loading well before the high water.

We arrived at Boston on August 4th earning £38.10s.0d. for the freight, but expenses came to no less than £9.13s.3d. owing mainly to pilotage fees. Running back light to Hull we loaded oil cake for King's Lynn, and then repeated the trip for another 161 tons. That second cargo was finally discharged on August 23rd. A short while later we were back to Hull again, loading more oil cake, for Wells instead of Lynn. We then had orders for Keadby, arriving on the 31st August and loading 152 tons of coal for Gravesend.

We then had the news of Hitler's troops invading Poland, and that momentous historical milestone signalled the extinguishing of all navigation lights around the coast.

The imminent threat of war, hostilities in which we would inevitably be involved, was brought home to us by the issue of a .303 Canadian Ross rifle, and .303 single barrel Lewis gun. The Lewis gun was mounted to the deck just aft of the main horse over on the port side of the engine room hatch. What I considered far more important than our armament was a new sprit and leeboard, but neither of these were forthcoming.

We arrived at Gravesend and were discharged by September 7th. We returned to the shipyard at Strood for a bit of a refit and clean up. That interval from our trading took longer than anticipated, in fact the best part of a month, for we were not ready to receive cargo again until October 8th.

During that time I had to say goodbye to my mate. That old but experienced man who had been with me so long and had served at sea during the first World War, had come to the conclusion that one war at sea was quite enough. He packed his bag and swallowed the anchor. I was not the only one to miss him around, because my wife had known him as long as she had known me, and the two boys had grown up with him about whenever they had sailed with us. That left me with an eighteen year old lad, comparatively inexperienced, as mate. It meant an increased amount of work and worry for me, but before long I found a much more experienced bargeman to join the ship.

Our first charter after the refit was to carry 150 tons of maize meal from the Albert Dock, London to Great Yarmouth. Although we had to ride out a heavy gale in the Thames Estuary we arrived at Great Yarmouth on October 16th. By then of course, I'd got quite used to riding out gales, but the wailing of air raid warnings which we heard while discharging at Great Yarmouth was a taste of a new fear to live with.

From Great Yarmouth we went light to Goole and then returned to Gravesend with coal which was discharged by October 31st. Although the war was only about six weeks old freight rates were already rising dramatically. The first example of that was our coal cargo from Goole. It was fixed at 9/8d per ton, a jump of 3/6d. Such were the spoils of war for us, though the risks were just as real and many barges ended up mined or lost through other enemy action. Some came to grief because navigation was so much more difficult without the lights to guide us.

From Gravesend we went over to the Tilbury Dock and loaded a cargo for Ipswich where we arrived on November 13th. It was not until the 17th that we were ready to leave, under orders to proceed to Boston light.

The .303 Canadian Ross rifle issued to *Scone* was designed in the late 1800's. During WWI serious defects which caused it to jam had disastrous consequences for those in the trenches. It was replaced by the Lee-Enfield. However, a few survived and were given to the British Home Guard and coastal vessels in WWII, though their vulnerability was not disclosed to those to whom they were issued! (Imperial War Museum)

Navigating without lights, we had to grope our way outside the Scroby Sands then shape our way to the north'ard. Having a light westerly wind I thought we could make our passage all right, despite the fact that we could no longer receive any weather forecast.

However we did make progress, for we were below Cromer when the wind suddenly flew into the north-east. I turned south on the port tack. We dropped the topsail and had just brailed the mainsail when a squall hit us. Down crashed the sprit. At the same time the barge had put a lot of weight on the leeward leeboard and that also broke, drifting astern on its pendant where it acted like a paravane[1]. So with foresail and mizzen, we headed back towards Great Yarmouth. We were in a bit of a mess, bearing in mind I only had the services of a young and inexperienced mate. In the dark and in effect short-handed there was little to be done until the gale eased. However, we kept off the land and by the first streaks of dawn we had the Cockle lightship well under our lee.

[1] Mine-sweeping equipment comprising pairs of cables streamed astern, the cable ends fitted with boards which steer the cables apart and underwater to sever the moorings of submerged mines.

As soon as it was light enough we hacked away at the broken leeboard chain until the board went clear. We then headed for Yarmouth piers, with the Naval patrol boat frantically signalling us to heave to. Those requests I was compelled to ignore, for the broken piece of sprit swinging about meant that I was having the greatest difficulty keeping the barge under control. I offered that explanation to the Naval Controller when we got into Great Yarmouth, but it seemed to cut little ice. I am sure he had little clue as to the ways of sailing ships.

That same rough night saw the motor ship River Witham, with whom we had an argument over loading at Orford some four years previously, go ashore off nearby Happisburgh[2]. We berthed and although very tired I phoned the owners to let them know what had happened. Having returned to the barge we tidied up the broken spar and tangle of rigging. After a good sleep I woke to find a fair no'therly wind blowing so, in accordance with instructions which I had received the previous evening from the owners, we left Great Yarmouth and set a course for Strood for repairs. To my surprise, one of the firm's Directors accepted the responsibility for the accident and apologised for not listening to my earlier complaints about the sprit and leeboard. For getting the barge into port unaided he also gave me £5 for my troubles. Both a new sprit and leeboard were fitted.

[2] The River Witham was finally got off after the War, but only to sink going into the Humber.

With our repairs completed we went up the Thames to load linseed for Ipswich. Our cargo of 157 tons was on board by November 29th but we did not arrive at our destination until December 4th. That was because the war had already brought a very great change in the pattern of our lives. We were forced to navigate slowly, ever on the look-out for mines; moreover we were supposed to sail only during the hours of daylight. There was also the constant threat of attack from the air, a danger which was particularly severe on the East Coast. We were very lucky, for although we frequently saw wrecked ships which had not been so fortunate, we escaped the strafing suffered by so many. On one occasion we saw the Harwich based Trinity House tender moments after she had been shot up by German planes, but we were left unscathed.

After discharging at Ipswich we went light to Boston where we loaded for the Thames. We were back at Gravesend on December 18th and had to wait before we could discharge until the spring tides on December 29th. That was the end of 1939 and an even grimmer year was to come.

CHAPTER 22

Orders
for the Clyde

"I made it clear that we would happily have sailed the barge to Australia, but to go anywhere I had to have our compass reading true."

The New Year of 1940 began with us carrying the usual cargoes for the Humber ports. Although the war was just a few months old, freight rates for our own run had already doubled to typically £136 with expenses touching £13.

After taking on cargo at Strood, we left the Medway on January 3rd to go up the Thames to visit Millwall and Erith before finally leaving to make our passage north. It was January 20th when we arrived at Hull and after discharging we loaded for King's Lynn, which we reached on the 25th.

It was a severe winter that year and we had found it bitterly cold at sea. On arriving at King's Lynn, we had to lie for a night alongside Boal Quay. It had been snowing hard. Even in port the coldness struck home; so cold was it that the strong ebb tide brought a considerable number of large blocks of ice down the river, which threatened to play havoc with the bows of our wooden barge. In those awful conditions we had to rig large fend-offs of rope and timber over the side to prevent the ice floes from cutting right into the hull.

We did not finish discharging at Lynn until the 29th January, when we left for Boston, arriving there on February 1st. The weather had been so severe that our coal cargo was not ready and we had to wait over two weeks, until the 16th, before they were able to load us, *Scone* becoming frozen in as we lay inside the dock.

We eventually arrived at Gravesend on the 24th and had unloaded by the 27th. That slow freight turned out to be the last of the many coal cargoes that we delivered to the wharf at Gravesend. We had carried well over 20,000 tons of coal there during the eight and a half years we had spent in that trade. It had been good business to have, giving us regular full cargoes at a fair rate, whilst many other barges struggled and failed to provide a living wage.

After leaving Gravesend we carried a cargo of linseed to Ipswich from the Surrey Dock and then returned to Purfleet where we loaded coal from the SS Queensland for Strood Oil Cake Mill. Leaving the Medway on March 19th we

The collier SS Queensland lay at Purfleet when discharging part of her coal cargo into the holds of the *Scone*.
(National Maritime Museum, Greenwich, London)

returned to London to load another 150 tons of linseed for Ipswich which we discharged by April 4th.

A third, fourth and fifth identical run followed to Ipswich, each trip taking us approximately the same time, and all with nothing to make them out of the ordinary. The last of these cargoes was unloaded at Ipswich on May 21st and was the last 'commercial' cargo we were to carry for four long years.

On leaving Ipswich we had orders to proceed to Grays to load cement for Great Yarmouth. On arrival at West Thurrock Cement Works I was acquainted with amended orders which instructed us to proceed to the Deptford Victualling Yard and report to the officer in charge. Our contribution to the war effort was about to take a dramatic turn. Hardly had we secured alongside before men were putting steel and concrete shuttering around our wheelhouse. To my great concern that put our previously reliable compass out by as much as five points! On airing my worries the only guidance I could get was a suggestion that we should call at Sheerness for adjusting.

By the time the work was finished we were already part loaded with stores intended for Dunkirk and had orders to bunker to capacity and report to Flag Officer, Sheerness. Immediately we arrived there I pursued the question of our compass, which resulted in a heated argument between the Naval authorities and myself. I made it clear that we would happily have sailed the barge to Australia if they wanted us to, but to go anywhere I had to have our compass reading true. That came as a bit of a surprise and I half suspected that they gave me little credit for the ability to navigate. To illustrate my concerns I produced *Scone's* deviation cards, showing that the compass had been meticulously adjusted every two years in order that it could be relied upon. Without a solution we were sent back to Gravesend for orders.

There followed for me a most exasperating and frustrating time as a victim of inefficient officialdom. After a wait at Gravesend we were ordered into the Tilbury Basin. Once inside we were ordered out again to go up to Cory's hoist in Gallions Reach. Still none the wiser we loaded coal out of a steamer at night and then came the order that took the cake; we were to proceed to Margate, with a pilot! There was a tug ordered to tow three of us barges. As I most certainly knew more about that shallow coast than most, that needled me more than a little. I made my views known at Tilbury and as a result we were allowed to proceed to Margate Roads on our own.

I had to put our compass on the hatches to get it to read accurately enough for navigation. We anchored off Margate to await orders. After a lot of waiting a destroyer steamed up to us and ordered us to proceed to Ramsgate. That was the time of the Dunkirk evacuation and the harbour was crowded with many other craft of all shapes and sizes. It was indeed fortunate that the weather was fine. After lying outside Ramsgate for two days we were ordered into the harbour to be greeted by "Where the hell have you been?" "Obeying Orders." was my reply, just that and no more. Our cargo was discharged into the coal hulk in the inner harbour. Before leaving Ramsgate we were detailed to pick up a disabled motor cruiser which we towed back to Sheerness. On our way the remains of a barrage balloon fell down on our deck during a thunder storm!

Telegrams : " TRANSPORTS, WESDO, LONDON."
Telephone : ABBEY 7711.

Any further communication on this
subject should be addressed to :—
The Director of Sea Transport,
(address as opposite)

and the following number quoted :—

S.T.N. 4B.

MINISTRY OF SHIPPING,

BERKELEY SQUARE HOUSE,

BERKELEY SQUARE,

LONDON, W.1.

SECRET.

m.v. "SCONE"

Instructions to Captain.

With reference to the arrangements made for the emergency
requisitioning of your vessel, the Captain is to be instructed to report
to the Officer-in-Charge, Supply and Reserve Depot, S.R. D. Jetty, Deptford,
to load stores for North France. On completion of loading he is to proceed
to Dover and report there to the Vice-Admiral in Charge for further
instructions.

2. Wages. The Ministry has agreed the following scale of wages for this
service :-

	£.	s.	d.	new scale agreed
Master	6	18	10	£8.8s.10d.
Mate	5	14	1	5.14s.1d.
Engineer	6	2	5	
Deckhand	4	17	8	4.17s.8d.
2nd Deckhand or Engineer	4	17	8	19. 0. 7

These are inclusive of War Risk Money, overtime, crew providing own food.
It is requested that you will notify the men accordingly and advise the
Department whether the crew is prepared to sign on on this basis. These
wages should be advanced to the men up to and including 1st June. Where
the rates of pay already being received do not vary considerably from the
above, it is suggested that there may be no need to disturb your present
arrangements.

3. Food. The Captain should be instructed to take provisions for a period
of at least seven days, and arrangements are being made for iron rations
to be put on the vessel by Army Stores at Deptford, these rations to be
opened only in the case of extreme emergency.

4. Bunkers. These should be stored to full capacity and, where necessary,
they may be loaded at Purfleet on the voyage outwards and debited to
Ministry account. Stocks of lubricating oil should also be taken on board.

5. Clearance. The vessel should be cleared on Forms Sale 10, Sale 25
and Sale 35 as proceeding with Army Stores for a foreign unknown destination.
If any difficulty should arise in this connection the Captain should be
instructed to apply to the nearest broker for assistance.

6. Inventory. A list of stores and equipment should be prepared prior
to sailing. (This does not apply where an on-survey has already been held).

E. W. WINTERSON.

for Director of Sea Transport.

The letter from the Director of Sea Transport at the Ministry of Shipping, marked SECRET, requisitioning *Scone* for war service. A master, mate, engineer, deckhand and 2nd deckhand or engineer were considered as the appropriate complement, plus a couple of Naval ratings to man the armament. A new scale of wage rates were agreed in May 1942. The engineer and 2nd deckhand have been dispensed with, the mate and deckhand have no increase, but the master's wage is increased by over 20%.
(Albert Bagshaw collection)

'Harry', wearing his jacket with medal ribbons from WWI, says farewell to Fluff at the door of his Northfleet home.
(Albert Bagshaw collection)

Our next orders came as a considerable surprise for they presented a completely new prospect for *Scone* and her crew. "You are to proceed to Greenock on the River Clyde in Scotland via the Caledonian Canal."

In order to prepare for that epic journey we went back to our shipyard for a refit and to ship our old bowsprit, for I did not fancy making such a long passage without full sailing gear. I decided that Fluff should not come with us to Scotland, especially as we were expecting additional crew. She was taken home and had no trouble settling in; she had come to know all the family during their summer trips on the *Scone*.

Whilst on the yard we also had to exchange boats. Our replacement was a much heavier one with steel buoyancy tanks built in. It was fitted with a mast and lug sail and other incidentals that we might have required in an emergency. In the days that followed I worked out a way to use our motor winch to raise the new boat. I took a heavy line from the winch via a snatch on the aft head ledge of the main hatch and on to the davit falls. That system also proved useful for setting the mainsail, taking the sheet via the snatch and back to the winch. What a boon that winch had proven to be over the years, for we had also used it for warping the barge to her berth and hoisting the topsail. We loaded about 100 tons of chalk rubbish as ballast. All was going well for our departure but I was still worried about getting our compass swung. For the voyage we had to have four crew including myself. Extra accommodation was required for the fourth hand and two Naval Ratings who shipped aboard to man the Lewis gun for the trip. The shipwrights cut a doorway in the fore bulkhead from the fo'c'sle and built a temporary bulkhead within the fore hold.

It took until mid summer 1940 before we were finally ready and, although nothing had been done about the compass, we left Strood for Southend where I again protested about the problem. That seemed to have the desired effect at last and we proceeded to Great Yarmouth where we had the compass adjusted. It was not quite as good as before, but it was reliable enough. Our orders were to run northward to Shields. However the weather prevented us from reaching the Tyne in one leg and we were forced to put into Middlesborough for a short stay while it abated. Nevertheless we made the Tyne at our next attempt.

We were to proceed from there to Aberdeen, but we stayed for a couple of days awaiting better weather. Studying those unknown waters on the charts, Aberdeen seemed a long distance off to me. Leaving at dawn, once outside the Tyne we shaped a course outside the Farne and Holy Islands, then went on down to the Firth of Forth, fetching inside May Island by dark. We had a strong fair wind and did not require our engine, for we were making about ten knots under sail alone. When we reached the Forth we had to slow her down so as not to be seen passing Aberdeen before daylight, breaching the wartime regulations. Dropping the head of the topsail and brailing the mainsail slowed her down a lot. At dawn we were off Stonehaven, where we put on full sail again.

We later gybed to enter the harbour of Aberdeen. It had been a wonderful run under sail and the war would have seemed a long way off had it not been for the Lewis gun and our Naval Ratings. Unfortunately our fourth hand turned out to be of use and left us whilst we lay overnight in Aberdeen. I decided that a replacement was not needed.

We again left at dawn, the all clear sounding as the last of the German bombers headed for home, their deadly cargo having been unleashed on the city. We rounded Ratray and Kinard Heads and sailed into the Moray Firth and on up to Inverness which stood at the entrance to the Caledonian Canal.

Once it was daylight we could safely enter the canal. We passed through the first lock, then climbed up through ten more to pass along the canal into lovely Loch Ness and up to Fort Augustus. The next day saw us through as far as the western entrance at Corpach, near Fort William. Here I enquired if we could pass through the Crinan Canal, so as to save the long passage round the Mull of Kintyre. After explaining our length and breadth I was told that we could, but we were required in the meantime to sail to Oban for our overnight stay.

The next morning a drifter was sent out to pilot us through the boom defenses and on down through the Kelkregan Sound to the entrance of the Crinan Canal. We passed through the first lock quite all right, but when we got into the second we found the gate couldn't close behind us! The Port Captain had miscalculated, so back we had to go and come out again, anchoring for the hours of darkness. Underway again next morning at dawn, we shaped a fresh course round the Mull of Kintyre. It was towards the end of July 1940 when we finally arrived at Greenock, having made the passage from the Thames in three weeks.

A stark contrast to the flat landscapes of East Anglia and Lincolnshire, snow capped Ben Nevis towers over Loch Linnhe, Corpach and the Caledonian Canal. The entrance lock to the canal is by the tall 'engine house' building to the right, where the jetty and lighthouse can be seen nestling behind some trees. A crane sits on the wharfside midway between the entrance and the next lock.
(Alex Gillespie collection)

149

The London & Rochester
Trading Company's auxiliary
sailing barge *Knowles*, which
was under 'Harry's' command
from May 1923 to April 1924,
seen here loaded on the Norfolk
Broads in 1938. She was lost
whilst on the Clyde, but not
through enemy action.
(National Maritime Museum,
Greenwich, London)

150

CHAPTER 23

Naval Supplies

"Another time we loaded Mr Winston Churchill's personal effects and took them to the RMS Queen Mary"

The presence of a full rigged Thames sailing barge north of the border was met with some surprise and *Scone* was the subject of great interest as she passed through the canal locks under motor power. There was even greater interest when we made our way up the River Clyde to Glasgow under full sail to discharge the 100 tons of chalk rubbish we had carried as ballast. The spectacle of those brick red sails spread from lofty spars was short lived, for our next duty was to dismantle all our sailing gear, retaining just our main mast and sprit. Uncertain of how long our gear might lay ashore, I took great care with the way in which I placed it into store, making sure it would stay dry and protected. At the same time it was stacked in a way that allowed the air to circulate, to stave off the rot which might otherwise play havoc with our canvas.

We were the very first of the London River craft to make the trip, but within weeks several other auxiliary barges as well as Horlock's *Resourceful* and Goldsmith's *Success*, which by that time had abandoned their sails and were pure motor craft, came to join us in those strange waters. Our owners' *Marie May* and *Alderman* came too, the latter run down and sunk shortly after arriving on the Clyde. My old charge, the *Knowles* joined the fleet and she too was lost whilst there. The *Cabby* was another from our firm. G.F.Sully sent the *Arcades*, *Raybel* and *Hydrogen*, Samuel West the *Lady Gwynfred*, *Leonard Piper* and *Saxon*. All their gear went ashore as well and it seemed that those in command of the Clyde Anchorages Emergency Port did not envisage a swift end to the conflict. That turned out to be the case, with many of the barges not returning to their home waters until peace came to Europe over four years later.

After discharging our ballast came a period of four weeks frustrating inactivity lying in the Old Harbour at Greenock waiting for instructions. It was a very different pace of life to that which had provided our livelihood when we were trading by the share. Our first job was to load huge five ton blocks of concrete which were to act as weights for anchoring the moorings of boom defenses. It reminded me of the times we had loaded huge boulders of Portland stone down Channel. That job lasted about a week and entailed us working alongside a steamer, the old Scillonian, which had served the inhabitants of the Scilly Isles from Penzance when in trade. She was lying at anchor at the Tail of the Bank, just below Greenock. It was not long before I nicknamed that ship HMS Sewer as it was a far more fitting name. She lay high out of the water and head to wind, with all the waste waters discharging overside from the latrines and lavatories, raining right down on us as we laid alongside.

Shortly after that distasteful start our first proper job came along, carrying Naval armaments for a time. That task continued until the proper Naval supply

The Isles of Scilly Steamship Company's first Scillonian, seen here on the rocks at St.Agnes. She was undamaged. Having ordered replacement tonnage just before the war, she was destined for Belgian breakers, but diverted for service on the Clyde. (Rodney Ward, Isles of Scilly Steamship Company Limited)

ships came to Greenock. We loaded on one occasion for the submarine depot ship HMS Forth, which was lying in the Holy Loch. On approaching the ship we hailed with the news that we had torpedoes for her. Their reply ordered us alongside amidships. I decided to hold off, for she had a nice teak companion ladder hung overside where heads were then appearing ready to take our mooring lines. As the ladder did not move, I requested the duty officer to have the gangway hauled up out of the way as I was anxious to avoid damaging it. I made the request again and again but without result. We rounded several times and were getting nearer to the ship all of the while. I was still waiting for some action when all of a sudden a gust of wind came from nowhere and set us driving alongside the ship, crushing the bottom of her ladder to pieces. That resulted in the duty officer coming to life at last and we were in the midst of exchanging a few heated words when the Captain made an appearance. Firmly but politely I pointed out that as a result of his officer's indifference towards us his gangway had been smashed.

Although we visited that ship on many subsequent occasions, I never again saw that particular duty officer. However I often saw the Captain during our many comings and goings, and we'd often pass the time of day together during our visits. Although the gangway incident was never mentioned again, I got the feeling that my version of events had prevailed.

On the subject of torpedoes, one of our many jobs at that time was to take 'rogue' torpedoes to Arochor, which was the torpedo range at the top of Loch Long. Other than that work our main task was to carry ammunition of all types to all manner of ships. We delivered a wide variety of cordite, bombs, shells and depth charges before finally being taken off those runs and given the job of carrying general Naval stores; a seemingly safer occupation.

For some time I had been looking to bring the family to Scotland. They had been living underground at night for three months because of the latest bombing, and before that it had been the same during the Battle of Britain. The home built concrete air raid shelter in the back garden was pretty awful. It had water running down the walls and was not a nice place to spend so much time. So they locked up and left for Greenock; Kathleen, Cyril, Albert and Fluff the dog. They were bound for accommodation I had found in Forsyth Street. The house was on the top of a long road which ran up hill from the River Clyde. It was a large house with an 'artists' room on the roof with spectacular views of both the snow covered mountains and the great ships wearing their drab wartime grey. The house was owned by an elderly Mrs. Ramsey who lived there with a much younger lady, Mrs. Ferguson. They had a Scottish Terrier called Jock whose presence dictated that Fluff should return aboard the barge.

The boys got started in a local school, but soon found it hard going. Discipline for misdemeanours was enforced by a three tailed leather strap, either to individuals, or the whole class if the culprit was not revealed. There was resentment shown to those who had come from south of the border. So much so that the headmaster eventually told all the pupils at assembly that my children had come north to escape Hitler's devastating bombing and did not expect to face fresh hostilities from their own people. Apparently things were much better at school after those few words.

The Greenock Harbour Trust issued permits to those authorised to use the dock facilities.
(Albert Bagshaw collection)

One day the boys had a reminder of the bombing in the south when the air raid warning sounded at the end of their school dinner. Cyril was already back in class, but Albert's age group had a later break. Apparently he ran out of the main gates with the crowd, thinking they were on their way to the shelter. In fact there were no shelters, so the pupils were sent to their homes and young Albert was heading in the wrong direction. Luckily a lad who had befriended Albert spotted him and pointed him in the right direction for home. By the time he was nearly back he met up with his elder brother who had been sent by their worried mother to find him. In any case as it turned out there was no raid.

The large house enabled us to live separate lives from Mrs. Ramsey and Mrs. Ferguson. They did all the cooking and washing and I provided the coal for the fires. These were needed in every room for otherwise the place was bitterly cold. After a couple of months I thought arrangements were well settled, but then Mrs. Ferguson began to make things unpleasant, possibly resenting our intrusion at the house or perhaps our south of the border origins. Whatever the reason, I arranged for a move to some alternative accommodation just down the road.

We were often surprised at the lack of common sense in the way we were ordered to work. Maybe it was the Navy's way in wartime but it was lunacy to me. On one occasion we had been lying in the Albert Dock at Greenock at a time of particularly foul weather. It was mid-winter time and that type of weather was to be expected, but what was not expected were orders in those conditions to take a small box out to HMS Titania lying in the Holy Loch. Owing to a strong easterly gale blowing I at first refused to take the barge out, for we were too low powered and I was very doubtful of our performance once out in the loch, to the extent that I could see us ending up down to the leeward and in trouble.

It seemed ridiculous to choose us to go, or any barge. There were a number of high powered tugs available which could more easily have performed the task if it was really necessary. However, after several refusals a Navy Supply Officer named Stevens finally persuaded me to go. We waited until low water and then set out. The weather was absolutely foul. Nevertheless we eventually made our way to the ship and arrived with a heavy swell running which prevented us from getting anywhere near her. There was no lee side and after a number of unsuccessful attempts I decided that all we could do was to get a line across to her while we stood off, so the box could be dragged through the water and got aboard. With much difficulty that was what we eventually achieved.

Because there was a heavy sea running it was really impossible for us to punch our way back towards Greenock. I decided we had to go to the top of the Loch to find some shelter. Though far from comfortable we let go the anchor to await a chance to get back. It was some days later before the weather had improved and we were able to return to Greenock. Before long we were alongside the Titania again. I enquired about the box, as to whether it had been damaged in the hazardous business of getting it on board. It was then that I learned it contained delicate medical instruments for Malta. It seemed ridiculous to me to have them sent aboard in such bad conditions in our unsuitable craft, especially was the Titania did not leave until some days later.

My mate, William Devenish, was approaching retirement age and had not been enjoying good health. He decided to quit and return south, but not before he had asked me if his son Larry could fill his shoes. Larry Devenish had been on convoy work where he had already been torpedoed two or three times. The relative safety of the barge work on the Clyde must have been a prospect he viewed with some relief. So father and son changed places aboard *Scone*, William leaving the sea forever.

Many and varied were the cargoes we carried in those days. On one occasion we took out a quarter of a million pounds in boxes to the BI[1] liner Narkunda. Another time we loaded Mr Winston Churchill's personal effects and took them to the RMS Queen Mary. We made many visits out to the anchorage off the Tail of the Bank, calling on both of the two big Queens in turn after they had brought over American troops. Our little part in that build up of fighting force was to take off some of those men together with their stores and rations and deliver them in Glasgow. During the run up, we had the opportunity to talk to some of these young men from the other side of the Atlantic. They were trying to come to grips with our type of money and enquired how things in general were really like over here. Asking one of them what kind of Atlantic crossing he'd had, he said he didn't realise

[1]*British India Steam Navigation Company Limited.*

Even the wartime 'drab' paintwork of the Queen Mary, operating as a fast troopship across the Atlantic, cannot disguise the magnificence of such an ocean thoroughbred. (Conway Photo Library)

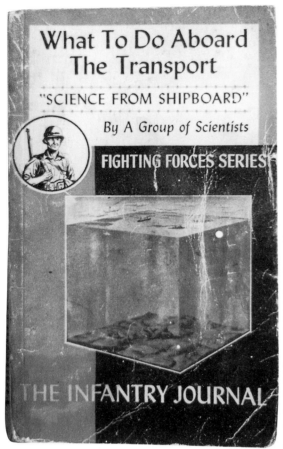

The 'guide book' issued to American troops on passage to Britain.
(Albert Bagshaw collection)

there was so much water in the world. For some of these men it would seem that water was something in a pond or tap, and ships were either in a bottle or a book; many of them came from far inland in their native country. To help them to understand that new experience they were issued with a little book that was called What To Do Aboard The Transport.

There were other ships that we visited, many of them whose names have since featured in epic stories of bravery during the war. One was the tanker Ohio which we loaded with stores before she made her famous dash to relieve Malta. The battleships Repulse and Prince of Wales also received stores from us. On reflection, I really believe that we must have been alongside practically every ship in the Royal Navy at that time, from the largest to the smallest. Names run off the tongue, other battleships included the Revenge, Rodney, Renown, Ramilles, Resolute, Nelson and Royal Sovereign.

Another particularly regular customer for us was the gunnery cruiser Dragon. There were also the cruisers Norfolk, Suffolk, Malaya, Exeter, Diomedes, Kenya, Cleopatra, Newcastle, Sheffield, Scylla, and Belfast, many of which were to be involved in memorable engagements which may well have changed the course of the war. Of course there were many other ships of every type including aircraft carriers, destroyers, frigates, corvettes, minesweepers and patrol craft. There was the American carrier Wasp, as well as the Free French Le Pasteur and a lot of other big liners besides the two Queens.

The Queen Elizabeth receives stores whilst anchored on the Clyde. She sports a large calibre gun aft and anti-aircraft batteries on her upper decks, though her best defence against the enemy was probably her speed.
(Photo: P Ransome-Wallis, Bob Ratcliffe collection)

Aboard one destroyer we came across was an officer with scrambled egg on his peaked cap fishing over the side with rod and line. Fishing being my favourite pastime, I enquired "Had any luck, sir?" shortly after we were secured alongside. "Not so far" he replied. I though that a little strange, for fish were plentiful around those parts and his fishing tackle seemed far superior to what I was using with considerable success. "What bait are you using?" I enquired. It turned out to be something that his ships cook had found for him and was obviously no use. "What you need is some of these" I said, passing over some ragworm out of my bait box at the back of the wheelhouse. "Where can I get some more?" enquired the officer, and I told him that we dug them from the sand behind the Gourock pier at low water.

A few days later I had started to dig for more ragworm when I found a Naval Petty Officer complete with bucket and fork digging away without success. I wandered over and to his amazement named his ship! "My Commanding Officer sent me to dig for bait, but I'm not having much luck. How did you know of my ship?" he enquired. Passing a couple of handfuls of ragworms to the astonished officer I told him "Tell the Commander these are with the compliments of the *Scone*." and relieved that he would not return empty handed he thanked me and was on his way.

Our contribution to the war effort was to serve those ships. Some we found it a real pleasure to help while others were offhand and unhelpful to us. Generally speaking we found the larger the ship, the more trouble it was getting their ammunition and stores aboard. Sometimes the etiquette of seniority got us into trouble for not doing our job the Navy's way. We were carrying Naval stores to a number of ships including the depot ship HMS Cyclops in Rothesay Bay. We had loaded a large consignment of stores for that ship in the bottom of the hold, with some large boxes containing periscopes on deck. In addition we had several small lots of stores for other ships. Being none the wiser about Naval procedures, we set off on our rounds. Our course took us first to discharge our deck cargo, then all the other parcels until eventually just Cyclops' stores remained in the bottom of our holds. When we finally got to HMS Cyclops I was asked why I had left her until last? Furthermore, didn't I know that she was the senior ship in the port. I apologised and explained that all ships were senior to me, whether they were trawlers or battleships. I delivered my cargo as loaded so that it was discharged to the best advantage, without humping and shifting to gain access to cargo hidden beneath other cargo. I am pleased to say that the logic was acknowledged and we had no further complaints in that regard.

But we were once in trouble again because of a mixed up signal. That particular day we had been on the old coal run to Toward Point. We got back to Greenock about 4pm. Although we had our orders for an early start the next day we had a message to call in the office before leaving. We were told that after proceeding to Glasgow for another load of coal we were to take a ton of coal out to a destroyer on our route back. She was a brand new ship, having just been towed down from Glasgow where she would normally have bunkered with her oil. But for some reason or other the fuelling had been deferred. Having got to the ship we found out it did not have a coal fire aboard! Furthermore they had no means of getting any hot food so back we went to the harbour.

By that time all the working men had gone home for the day. I asked the two dock watchmen to give me six drums of diesel oil for the ship, which they refused to do without documentation and authority. I could understand that, but the poor devils off afloat had nothing, so I helped myself with the watchmen looking on. They went up in the air, one even threatened me with all sorts of aggravation as we rolled the drums along the quay and aboard the barge. But that made no impression on me and off we went to the ship. It must have been about 10pm by the time we reached her. Compared with the wrath we had left behind us, we were greeted with open arms and enthusiasm at the end of our journey. What ever was in store for us next day, we had made them happy that night. Next morning I was asked to report to the Stores Officer, fully expecting to exchange a few harsh words. But instead he thanked me for helping out, as the original signal had been misunderstood!

Things were getting a little better down in the south so we had discussed the idea of the family returning to Gravesend. Although everybody had been happy in their temporary home for well over a year, our landlady had come to realise that the influx of service personnel and their families meant her rooms could command a much higher rate than we were prepared to pay. So it was decided that Kathleen and the boys should move out and live aboard *Scone* for a few weeks before heading for home. For Albert and Cyril it could not have been better or more exciting, visiting the big liners and warships. When we went alongside one of the Queens you could see the amazement and disbelief at the size of these vessels, their flaking grey paint seeming to stretch forever skywards. Of course, these anchorages were in very deep water and once secured alongside out would come the fishing tackle. Not only was Albert surprised by the height of the ships side, but also by how much fishing line he needed to reach the bottom. By the time he'd lifted his rod a couple of times to feel if his weight had really landed, his line went tight as he hooked his first cod. For a twelve year old it was an exciting moment as he brought the fish on the deck. Though size was not important at that moment, no doubt he would be holding his hands some way apart as the story was retold in the years to come.

Despite wartime austerity the paperwork continued, *Scone's* Half-Yearly Return for the period ending 31st December 1942 was lodged at Greenock on 7th January 1943. (Albert Bagshaw collection)

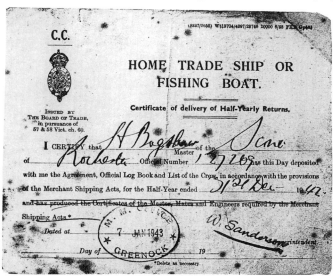

The boys had not forgotten the pleasure they had got from messing about in our old barge's boat before the war. But our new boat, with its heavy steel buoyancy tanks was not easy for young lads to row. Those Scottish waters in winter time were no playground either, with the squalls that rolled down from the mountains to plough the surface of the lochs. I had never taught Cyril and Albert to sail. I had to sail for a living and didn't need it for pleasure. But one day Larry was keen to show them how. We were anchored in the Holy Loch, so off they went with the lugsail hoisted, the breeze propelling them along in grand style. They took to it with enthusiasm, a new and exciting way to get about.

At the end of March the boys returned to Gravesend, leaving Fluff behind on the *Scone*. They had only been home a couple of weeks when the Luftwaffe bombed Greenock and its surroundings in earnest.

Perhaps less noteworthy than ammunition and torpedoes, but of equal importance were the cargoes of coal and coke we carried out from Rothesay Dock, Glasgow. We carried them to Dunoon, a shore base named HMS Brontosaurus at Coates Castle, Toward Point and to Strachur up Loch Fyne. We also visited Inverary, Lamlash, Brodick Bay, Cambelltown, Loch Ranza and Rothesay down through the Kyles of Bute. Carrying coal cargoes once more seemed like old times to us, although at some destinations, like Loch Ranza, the methods of unloading were very different. The first time we arrived there we were put alongside a small pier which was not really a safe berth. On top of that problem, when the Sergeant Major of the Marines arrived he informed me that his Jeep was all the transport he had to cart away the coal! As we had done at Gravesend, we hove it out of the hold by the use of our motor winch, only to wait ages for it to be carted away.

While the unloading slowly proceeded I realised that our berth was open to the whole drift of Loch Fyne. Seeing an old tower at the end of a shingle beach, I was soon stepping it out ashore to have a look to see if anywhere would fit our purpose. I was pleased to find a sheltered shelving beach beyond the spit where *Scone* could safely dry out and unload. Then I had to sell the idea to the Sergeant Major. When he returned I asked him how many sacks could he could lay his hands on. He said he would look, but first wanted to know why. I asked him to accompany me to the spot I had found. "How would you like me to dump the coal here, high up on the shingle shore. You could have it collected as and when you need it." It was good that he was a practical man, for he didn't hesitate. We moved the barge so they could unload us when the tide went out.

Next morning the Sergeant Major arrived with a good group of men and asked for my plan. He had found plenty of sacks so I asked him to split his team in two. We'll have half of them down below in the hold" I said "and the others on the beach." Down in the hold they filled the sacks with as much as a man could carry. We hove them up on the winch and lowered them over the side. The other men carried them ashore and emptied the coal on a growing pile, returning the empty sacks for re-filling. After an hour or so the men changed places to share the work around fairly. They worked well and the Sergeant Major was pleased. So was I, for we finished next day quite early.

It as a pleasure to help men like that, but it was not always so simple. On another occasion when we had troops to unload our coal there seemed very little enthusiasm for the task. The Tommies worked so slowly I thought we would lay on the berth forever. As the squad was not under my authority I had to tread lightly. In a round about way I tackled one of the men, who informed me that at the end of each day they had to get rid of all the coal dirt and turn out on drill parade for inspection. I took the NCO to one side and put forward a suggestion that the men should be encouraged to unload us by an offer of recreation time at the finish of the job, and no parades. Somewhat to my surprise he agreed and our barge was empty in no time! I had the feeling that after that trip they looked out for *Scone* in the hope we would return in the future, which from time to time we did.

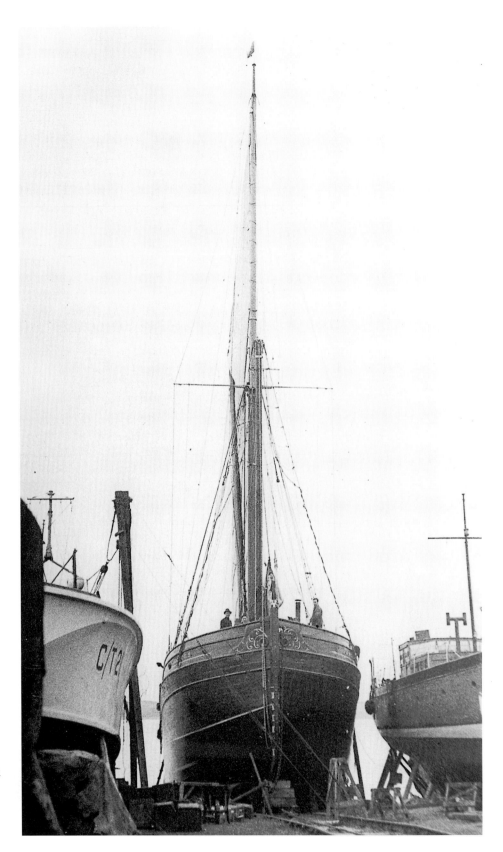

Sully's *Arcades,* built in 1921 as the *Olive Mary* at the Sittingbourne yard of Wills & Packham Ltd, fits out for her passage back to the Thames on Fife's slipway at Fairlie. She was one of the largest wooden spritsail barges ever built. She ended her days just a few years later when burnt to the waterline due to a fire in her cargo of baled straw, whilst on passage from Colchester to Ridham Dock, Kent, in June 1947.
(Scottish Maritime Museum, Irvine)

CHAPTER 24

The Return Home

"... Scone was the only Thames sailing barge that ever brought a cargo from the West Coast of Scotland"

Towards the end of our stay on the Clyde we sometimes carried mail and general cargoes to Glasgow and Paisley as our more warlike roles declined. With the war in Europe in its final chapter, I had steadily been going over our rigging and sails in storage ashore with every opportunity that came my way, ready for the day we would return south. By the time we received orders from our owners to return to London in mid summer of 1944, *Scone* was ready for fitting out. When I enquired of my superior at Greenock about our release, I was told to start getting our gear ready. Much to his surprise I told him that the task was already accomplished. He promised to do the necessary paperwork for our release within a day or so. Before setting out we loaded 100 tons of lead at Glasgow and then lay there for a while fitting out for the passage south.

It was midday when we left Greenock for the last time, feeling good to be like a sailing barge once more with all our gear up aloft. It reminded me of the first time we were working out to the aircraft carriers when we could not draw the deck's attention to our presence. They were so high up that the mate had to climb to our masthead to communicate.

The first leg of the voyage south was to Oban by way of the Mull of Kintyre and through the Sound of Islay. We had a dirty passage with visibility obscured by heavy rain. That made things most unpleasant for us as it was a bad coast to navigate, even in good visibility. However we made Oban first stop, and in such weather I felt we had made a good passage of it. The following day saw us make the run from Oban to Corpach near Fort William, at the western entrance to the Caledonian Canal. There was quite a strong

The *Leonard Piper* and *Hydrogen* re-rig their sailing gear in the Kingston Dock, Glasgow for the return south, having spent their time on the Clyde as pure motor barges. The sails appear to be undressed and probably new; perhaps their old suits had fallen victim to poor storage. Note the large boards on the front of each wheelhouse bearing the barge's name. (Glasgow Museums, Museum of Transport)

Arcades crew are for'ard seizing a new foresail to the hanks, whilst her new mainsail and topsail are set, no doubt so they may begin to stretch to fit her spars and rig.
(Scottish Maritime Museum, Irvine)

Her sails stowed and leeboards re-shipped, *Arcades* is seen about to float from Fife's slipway ready to start her long journey home.
(Scottish Maritime Museum, Irvine)

breeze blowing, and with the windage of our high gear aloft, the effect of which I had been without for some years, I had some job to take the way off her, even with the engine in reverse. After spending the night there *Scone* entered the canal and passed on through to Fort Augustus in front of a westerly gale. Believe it or not, the day we reached Loch Ness the wind came due east. Making enquiries when we arrived at Inverness I was told the wind had been easterly for weeks. That meant that we had to wait for suitable weather and it was several days before we could put to sea and make Aberdeen, our next planned port of call. From Aberdeen our next stop was the Tyne. Then it was the old familiar run to Great Yarmouth and on to Gravesend.

We had some trouble just before our long passage ended. We were forced to spend the last night at anchor in the Barrow Deep. When we went to get under way the splice on the stanliff began to draw after we had set the mainsail. We managed to put the runner tackle on the heel to temporarily take the strain and to prevent it from drawing further. If the splice had failed our sprit would have come crashing down on deck, causing no end of damage, if not injuries. We managed to get to the Lower Hope, where we anchored and had a good night's sleep, being physically and mentally exhausted after a passage not far short of 1,000 miles.

The next morning, with the last part of our journey covered with great care we went alongside Lovell's Wharf at Greenwich to discharge. In all, our run home had taken just short of three weeks. I believe that *Scone* was the only Thames sailing barge that ever brought a cargo from the West Coast of Scotland, for the others on war duty came home light.

From Greenwich we made our way down and round to Strood for a complete refit. It was a much needed one after all the time we had been working, practically without respite, and so it took quite some time. She had all new rigging to replace that which I put on her in 1925, 19 years before. A lot of water had been under her since then. *Scone* also had a new keelson fitted, put in with all new keel bolts. In fact anything that needed doing was done and she came off the yard looking like a new ship.

It was during our time on the yard that the company offered me a new 750 ton motorship, but I turned the offer down. I told them I was thinking of swallowing the anchor and getting a job ashore, if one was in the offing.

By November 2nd 1944, we were ready for work again and our first charter was to take 174 tons of linseed to Ipswich from the Royal Albert Dock. We finished loading by the 9th. With the short winter days the light faded early. That meant us having to anchor up by the Whitaker for the night, near the Shore Ends in the River Crouch, as there were still no lights with which to navigate. To make things more unpleasant there was a continuous stream of Flying Bombs passing over our heads. Added to that was the noise of the heavy gun fire which was having a go at them. As we watched it was clear that the guns were having some success, but most of these wicked weapons got through to wreak havoc around the country.

Arriving at Ipswich on the 11th, we stayed a week before returning south with linseed cobbles for Silvertown. Then we went to Strood and loaded again for Ipswich, arriving on November 28th. Once empty we loaded flour for Strood, arriving on the 7th December. We were like a shuttlecock between those two ports, for at Strood we picked up 140 tons of cargo for Ipswich and we were back there on December 14th. Once again we were to return to the Medway, that time with 162 tons of oil cake which we loaded between the 19th and 21st and which was discharged after the Christmas holiday.

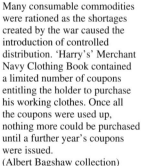

Many consumable commodities were rationed as the shortages created by the war caused the introduction of controlled distribution. 'Harry's' Merchant Navy Clothing Book contained a limited number of coupons entitling the holder to purchase his working clothes. Once all the coupons were used up, nothing more could be purchased until a further year's coupons were issued.
(Albert Bagshaw collection)

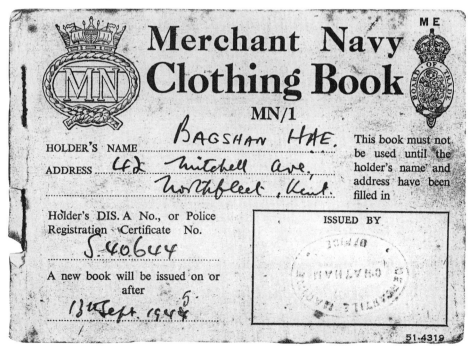

CHAPTER 25

**Farewell
to the Sea**

"... we arrived alongside the ship, only to be informed that we were not wanted after all."

The return south from Scottish waters was welcome in more ways than one. During the time we had been up on the Clyde we had been working for a weekly wage. Now that we were back, we were trading on more like pre-war commercial terms. Although some of our old regular cargoes were not around, nor were ever likely to be again, trading was good. Freight rates were higher than ever before. They were averaging out at about £90 per trip; many were over the £100 mark.

Our first voyage of 1945 was from the Victoria Dock Silo with 148 tons of linseed for Ipswich. We returned with wheat which we discharged at Silvertown. On January 19th we were back in the Victoria Docks again loading linseed for Ipswich. Moving from our discharging berth in Ipswich Dock we took on oil cake for Strood. We locked out a little after high water, carried the ebb from the Orwell and the flood up the Essex coast towards our destination, where we arrived on January 28th. We were not able to get our cargo discharged until February 6th, which resulted in the barge coming into demurrage, bringing the total freight to well over £100.

More and more linseed followed, the next consignment from Millwall Dock to Ipswich. We arrived back in the Ipswich River on February 13th and were at last discharged by the 17th. We had to return light that time to the Surrey Dock to load again. That was a similar cargo of about 155 tons and was discharged at Ipswich by the 24th. Returning light again we loaded 2,046 bags of linseed at Chamber's Wharf, Rotherhithe. That trip we were lucky for we found a cargo of 165 tons of flour to carry on our return to London.

The pattern continued; linseed down to Ipswich and back up with flour. At times we were right up under the London Bridges, which meant of course the additional work of lowering our gear. Sometimes we loaded in the Pool, as we did on April 11th when we took aboard 145 tons of linseed at New Fresh Wharf for Ipswich. But the routine was broken at the end of the month when we were chartered to carry nearly 3,000 bags of potatoes from the SS Macclesfield, loading in the West India Dock for delivery to Strood. After discharging the potatoes it was linseed again, loading 155 tons from the SS Mantold for Ipswich. Before we sailed I received additional orders to get the barge to Ipswich just as fast as possible. Such an instruction was a bit of an affront to me, for we made every passage just as quickly as we could. Apparently the mill was short of seed and would probably have to close down if our urgently awaited cargo didn't arrive in time. That meant we had to leave the dock and get down river beyond the Southend boom before dark and avoid approaching Harwich before daylight. We had to find our way from Southend to Harwich in the dark with no lights to help us; our way was between the maze of sand banks which were made more dangerous by wrecks and mines.

Wrecks littered the Thames Estuary, the most dangerous lurking just beneath the surface. Many barges fell victim to such obstacles or the contact mines which broke adrift from their moorings. The destroyer, HMS Gypsy, mined early in the war, was typical of the kind of hazard to be encountered when under way at night or in poor visibility without the benefit of lights. Note how the horizon in this Port of London Authority wartime photograph has been crudely painted out to prevent identification of the location. (Courtesy of Museum in Docklands, PLA collection)

After passing the edge of the Shoebury Sand we carefully felt our way round the Maplin using the lead line, picking up the South West Middle and then the Spitway buoy. There we had the choice of going over the Spitway between the Buxey and the Gunfleet Sands or to go down round the Gunfleet. With the tide the way it was I chose the Gunfleet and it was a case of dead reckoning all the way.

Finding the sands and edging cautiously round took us to somewhere on the edge of the Cork Sand by daylight, from where we headed into Harwich. It was the 16th May when we arrived and after discharging on the 17th we had to make our way back in similar difficult circumstances. It was then a question of getting out of Harwich by dark and reaching Southend by daylight.

We continued to trade to and from Ipswich, taking seed or cake into the port and often returning to Strood with flour. I found the very slow turn round increasingly frustrating. Following our arrival at Ipswich on May 3rd, we were still there on June 14th. After that passage we took flour back to Strood then returned with yet more linseed. We arrived back at Ipswich on June 26th and lay about until July 5th waiting to unload. Those delays badly affecting our earnings, despite the good rates and demurrage. Returning to the Surrey Dock light, we were locked in to load timber for Faversham, but a dock strike prevented us getting our cargo until August 3rd. More and more we seemed at the mercy of others.

We then had to go to Ipswich light to take linseed out of a ship laying in the Orwell and take it up to the mill. Following that we loaded oil cake for Mark Brown's Wharf in the Pool of London, just through Tower Bridge. Unfortunately we had to lower down as the bridge was out of action. With our cargo ashore we went over to Cole's Wharf, and carried away 96 tons of ground nuts for Strood.

Our next orders were to load bagged oil cake which had to be delivered to a ship in the Regent's Canal Dock. By the time of our arrival the ship had not made an appearance. After some debate we were told to move out and go into the Millwall Dock. Our cargo was consigned to a port in Norway and I offered to take it there direct in the *Scone*. Indeed we had almost cleared to go when a ship became available to take the freight.

Our next charter took us up under London Bridge to load bagged ground nuts. That made me absolutely furious, giving us the extra work of having to lower our gear to get under the bridge, when at the same time our company's unrigged motor barges were loading below the bridges. However, we had no say in the matter, so we went up and loaded 1,948 sacks, leaving on September 27th and arriving at Strood on the following day. From there we carried 150 tons of bagged oil cake to Ipswich and then loaded a return cargo which we brought back to Northfleet to await orders. We lay about laden at Northfleet until October 14th when we received orders for Strood, where we were not able to discharge until a full two weeks later.

While we lay at Strood I had taken the opportunity of visiting my home, an all too infrequent pleasure for those who earned their living at sea. Returning by road I met our London runner, the man who looked after arrangements for the firm's craft in the docks. I asked him when we should next be wanted. He told me we would be going to the Albert Dock for the mornings work, adding for some reason that I was not to tell the office that I had seen him. Sure enough we received our orders to sail that night. At 8am on the following morning we arrived alongside the ship, only to be informed that we were not wanted after all.

The same man who I had seen the previous day then came along and told us to go up to the West India Dock. That annoyed me very much, for not only had that nonsense cost a night away from home, but also we had lost our sleep for nothing. While we lay in the West India Dock the same runner came with another gentleman who was introduced as having been promoted in the firm to the shore job I had wanted. That was unbelievable. We were loading for Strood. As soon as we arrived I went straight to the company's office to see the Director. There it was confirmed that, despite my previously advised interest in the job, the man I had met had been appointed to the position.

When he told me that I had not even been considered for the vacancy, I bluntly told him I was leaving. When he protested that I could not just walk out after the twenty-seven years I had served with the firm, my simple reply was that my belongings were packed and all I needed were my cards.

And so on the 5th November 1945 I left the barge and the sea for ever. I had done 21 years and 9 months in the *Scone* and had spent almost thirty years in sailing barges. Together we had traded from the Thames and Medway to Penzance in the west, to Dieppe on the French coast, up beyond Antwerp, and north to the west coast of Scotland.

I had come to the conclusion that the responsibility was not worth the worry. With inexperienced crews and the income tax, the long hours and the increasing limitations imposed on our livelihood by others, it was just not worth the effort anymore. It was generally thought and often said that one prospered by being honest, trustworthy and hard working. From my own experience, I had nothing but doubt as to the truth of that. Certainly what I had achieved by my endeavours along those lines seemed little if any better than the spoils of wasters and cheats.

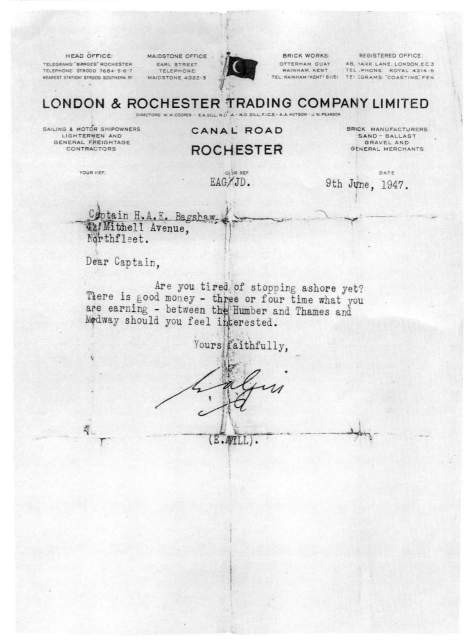

Economical with his words, E.A.Gill writes to 'Harry' with an offer of re-employment. (Albert Bagshaw collection)

A couple of years later that same Director wrote a personal letter to me asking if I would be interested in returning, saying a good living could be had between the Thames and the Humber. But I politely refused, adding that I had refused other offers to go elsewhere, where vessels of up to 2,000 tons had been offered me. My days at sea were over for good.

POSTSCRIPT

It is very clear from the closing lines of Captain Bagshaw's narrative that he was 'swallowing the anchor' a disillusioned man. He was just 44, but had already spent nearly 30 years afloat and many would have expected him to stay at sea until retirement. His disillusionment was borne of a variety of factors. He felt let down by his employers, who had always been slow to respond to his innovative ideas and had apparently failed to take seriously his request for a job ashore. His desire to give up his life afloat was not just in order to spend more time at home with his family. It was more to do with the trouble and strife that was clearly beginning to take over in docks around the coast and particularly in London. Restrictive practices, demarcation disputes and go slows all fuelled by an organised, powerful, overmanned and volatile dock labour force were beginning to tear the heart out of the world's busiest port. The frustrations of saving a tide here and a good swift passage there, only to see the gains wasted by the deliberate delaying tactics of the dockers was too much to bear, not only in 'Harry' Bagshaw's pocket, but also in a mind which had always held faith with a 'working hard hurts nobody' philosophy.

His employment took several turns after leaving the barge before finally he became berthing master at a large paper mill wharf on the south bank of the Thames, close to his Northfleet, Gravesend home. He had the good fortune of never needing time off through illness for the whole of his working life, though his younger son recalls one bout of influenza, to which he did not succumb.

The author's affection for sailing barges never left him, his interest exemplified by the notes he made which formed the basis of this book, and a further unpublished manuscript recalling in detail the operation of de-rigging and fitting out *Scone* in 1925, which he had completed just before he passed away in 1980 after a short illness.

And what of the craft themselves. The once familiar sight of a Thames and Medway spritsail barge in trade finally disappeared in 1970 when Cambria passed to the Maritime Trust. Many of these old 'Sailormen', as both the craft and the men who manned them were known, became motor barges, before variously ending their days broken up or

Scone, sails forsaken, motors deep laden from the Royal Docks on 24th April 1969. (C C Beazley)

168

A target for graffiti and vandalism, *Scone* lay at Twydall abandoned awaiting her fate. (Albert Bagshaw collection)

abandoned; but some of them were restored to sail by enthusiasts as yachts and charter craft.

Scone was occasionally seen by 'Harry' from his riverside office window, as she went about her trade up and down the Thames. Her main sailing gear was not stripped out until 1957 from when she plied her trade under engine power alone. It must have been about 1972 when 'Harry' re-boarded her for the first time. She was by then privately owned and laying at a marina on the Medway. The engine was started by the new owner, much to the delight of her old skipper, and the intervening decades just melted away as he recalled his days in command.

That was more than could be said for the rest of her. The rails were missing from the middle section, the sail horses had been removed. She had just a small pole mast placed right up near the fo'c'sle hatch and to complete that new guise there was a steel wheelhouse and steel hatch coamings with their angle iron bracing.

Some months later *Scone* moved down river to a marsh mooring near Twydall, Gillingham. There she remained for about three years, during which time she was driven higher and higher up onto the saltings by the wind and an exceptionally high tide. That's where she stayed, drying out for month after month. The local children made her black hull the target for graffiti, and she was vandal stricken. In fact, it looked as if *Scone's* days were numbered.

By about 1976 rescue was at hand; a new owner to save the old work-horse. His name was Steve Mallett. Steve had served aboard *Scone* as a lad when he was 17. He had left her to go to the open sea in larger vessels. But he remained fascinated by her and then, aged 28, he was back as the owner of an almost derelict and rotting wreck. First the engine didn't work; it took a week of hand swinging before he could get it to fire up.

Next came the task of getting *Scone* back afloat. Steve, with the aid of a friend, dug a 40 foot channel through the mud and got her out. She then had a new mooring across the river by the Hoo Marina. Steve worked as a ship's Captain all winter, trading along the Spanish coast. Over the next few years he and his girlfriend Sara spent all their spare time restoring the barge back to sail.

Captain 'Harry' Bagshaw (right) with Tony Farnham and Owen Emerson (left), on board Owen's sailing barge *Victor*, seen here in the Gravesend coal berth so frequently occupied by 'Harry's' *Scone* some fifty years earlier. The old wharfside buildings have given way to public riverside gardens but St.Andrews Waterside Mission survives as an art centre. It was not long after this nostalgic moment that 'Harry' passed away.
(Kentish Times Newspapers)

By the summer of 1978, *Scone* was beginning to look the part once more. It was towards the end of that summer that the old skipper visited his former charge for the last time. Although she did get a brief outing that autumn, it was the following year that *Scone* really sailed again and took part in the annual barge matches.

Although she was under way again, it was not the end of Steve and Sara's task, for keeping the old timbered hull in good condition was to occupy much of their time in the years ahead. The starboard bow rail was laying on the deck when spotted by one of 'Harry' Bagshaw's sons, its place taken by new timber skillfully replacing the old. The visit was short but memorable, for when about to leave Albert nodded towards the old piece of bow rail with the barge's name cut in, still carrying the scars from the collision with the tug Crusader back in 1930. "Take good care of that." he said. "Oh I will." said the younger man. Just before Albert reached the shore, contemplating memories of trips aboard the barge as a child, there was a shout from behind. Turning, there was Steve struggling along the gangplank with the heavy piece of rail, drift bolts and all. "Here, you take it." said Steve, "Nobody deserves it more than you do." That much travelled piece of bow rail now remains with the family of the late skipper along with their kind thoughts of Steve and Sara.

It was in 1982 that they parted with the *Scone*. When the news of her sale got back to 'Harry's' family, a visit to Hoo found her all snugged down,

Under sail again for the first time in over twenty years, *Scone* is seen at the Swale Barge Match on Saturday 18th August 1979.
(Rick Hogben)

but bound for where, they did not know. Then one day the old berth she had used was empty, the barge had gone. From then until 1990 her location was a mystery to the Bagshaws. In the autumn of that year one of the family's relations happened to ask if they knew that Scone was laying in the Millwall Dock and being used as a floating restaurant.

Just a few days later 'Harry's' younger son was standing alongside his childhood second home. He thought he could hear voices down below. The sound seemed to be coming from a new entrance to the main hold by the main horse. Trying to think of the right words to say, he walked up and down the quay several times. Still there was no sign of life on deck. Finally he called out "*Scone* ahoy." Then again, even louder. A woman's head appeared out of the companion hatchway. "Good morning. Is the owner about please?" "Yes, I'm the owner." was her rather curt response. In the following seconds her stern face changed to show surprise and delight as he explained his interest, for she was about to learn something of the past history of her barge.

Her new role has *Scone's* hold full of comfortable furnishings and maritime memorabilia. (Diana Craigie)

Down below in the hold, where as a boy he had played many games with his brother, was a delightful dining room. The lady owner apologised for the way in which she'd greeted him up top and went on to explain that members of the public were forever wandering aboard the barge thinking it was a kind of museum piece. *Scone's* new owners turned out to be Clare and David Hunter. In the two hours that followed, the old skipper's son told much about *Scone's* past. Those folk were the second new owners since Steve sold her and in the years which followed that meeting she has changed hands yet again, now belonging to photographer Diana Craigie.

There must be a fear that in her somewhat sedentary semi-retirement *Scone* is at risk in a way that has overtaken the *Oak* and *Lady Gwynfred*, both of which effectively ended their days at dockland moorings after long periods of inactivity and with little but superficial repair. But that need not

The steel barge *Wyvenhoe* was built a century ago and was restored to sail after trading as a motor barge for over fifty years. She is seen here during the Blackwater Sailing Barge Match in the mid 1980s.
(Keith Yuill)

be the case and it is to be hoped that she may emulate the *Marjorie*, now actively under sail again after many years as a static hospitality venue for the London Docklands Development Corporation.

Of around 1,500 barges that plied their trade when 'Harry' Bagshaw first stepped aboard the *Gwynhelen*, just a couple of dozen or so survive in commission in 1998. It must therefore be some coincidence that one of these survivors is the 1898 steel barge *Wyvenhoe*, in which 'Harry' served in the early years of his career afloat. After being converted to a motor barge in the mid twenties she traded on until 1982 before being restored to sail.

The passing years continue to take their toll of these historic craft. The number of suitable hulls for restoration has dwindled dramatically and the majority of the few that could be considered worthy have already been re-rigged once in the past before falling on hard times. Costs for repair to keep pace with the inevitable deterioration that occurs with the passage of time continue to escalate to prohibitive levels.

But such a jewel as the survivors of that once great fleet represent in the nation's maritime heritage must not be allowed to diminish unnoticed until just one or two, if any craft remain. The barge matches, the origins of which go back over 150 years and which are such a feature of the Thames and Medway estuaries, would be sorely missed. No longer would we be able to see the spectacle of upwards of 50,000 square feet of tan coloured canvas spread, as a dozen or more barges converge on the start line at the annual races.

A little over a decade ago the *British Empire* was a fine example of a Thames sailing barge, restored to sail after more than half a century in trade. Now she lies derelict at Battlesbridge, with no realistic chance of survival.
(Keith Yuill)

If nothing is done such an outcome is almost inevitable. Let there be no doubt about it, these humble craft have played their part in peace and war in shaping the history of the nation; now is the time for the nation to do its duty in return - before it is too late.

Scone alongside as a static hospitality venue in the Millwall Dock, a place she used to visit for very different cargoes than those which come aboard today. (Diana Craigie)

INDEX OF VESSELS

Barge names are shown in *italics*.
Bold page numbers refer to illustrations.
Light vessels are listed within the **GENERAL INDEX** which follows.

GENERAL INDEX

The Thames Sailing Barge

SCONE

Drawn from the picture which appears on page 128. She lays starboard side to the wharf with her fore and main hatches open. To keep the hatchways clear, her furled foresail is hove to port by the 'bowline', the sprit pulled clear by one of her 'rolling vangs' led for'ard and her 'mainsheet' hung below the folds of the 'brailed' mainsail. The hatch covers are stacked for'ard of the fore hatch and fore and aft of the main hatch.

In the pursuit of clarity, some detail is intentionally omitted.

Drawing by Brian Young

Key to Drawing

1 Anchor windlass
2 Anchor windlass bitts
3 Bowline
4 Crab winch
5 Crosstrees
6 Davit falls
7 Flag, or 'Bob'
8 Fo'c'sle hatch
9 Fo'c'sle stove chimney
10 Forehorse
11 Foresail (furled)
12 Foresail halyard
13 Foresail hanks
14 Forestay
15 Hatch covers
16 Headledge (aft, of fore hold)
17 Leeboard
18 Lowers (brails)
19 Main brails
20 Mainhorse
21 Mainmast
22 Main runner
23 Mainsail (brailed)
24 Mainsheet
25 Main shrouds
26 Mastcase
27 Middles (brails)
28 Mizzen mast
29 Mizzen sail (brailed)
30 Mizzen sprit
31 Ratlines
32 Rolling vang
33 Sprit
34 Stanliff (standing lift)
35 Stayfall (or stem) blocks
36 Staysail (lowered)
37 Topmast
38 Topmast forestay
39 Topmast running backstay
40 Topmast standing backstay
41 Topping lift
42 Topsail (lowered)
43 Topsail sheet
44 Vang
45 Yard tackle